STATA MULTIPLE-IMPUTATION REFERENCE MANUAL
RELEASE 11

A Stata Press Publication
StataCorp LP
College Station, Texas

Copyright © 1985–2009 by StataCorp LP
All rights reserved
Version 11

Published by Stata Press, 4905 Lakeway Drive, College Station, Texas 77845
Typeset in TeX
Printed in the United States of America
10 9 8 7 6 5 4 3 2 1

ISBN-10: 1-59718-070-X
ISBN-13: 978-1-59718-070-2

This manual is protected by copyright. All rights are reserved. No part of this manual may be reproduced, stored in a retrieval system, or transcribed, in any form or by any means—electronic, mechanical, photocopy, recording, or otherwise—without the prior written permission of StataCorp LP unless permitted by the license granted to you by StataCorp LP to use the software and documentation. No license, express or implied, by estoppel or otherwise, to any intellectual property rights is granted by this document.

StataCorp provides this manual "as is" without warranty of any kind, either expressed or implied, including, but not limited to, the implied warranties of merchantability and fitness for a particular purpose. StataCorp may make improvements and/or changes in the product(s) and the program(s) described in this manual at any time and without notice.

The software described in this manual is furnished under a license agreement or nondisclosure agreement. The software may be copied only in accordance with the terms of the agreement. It is against the law to copy the software onto DVD, CD, disk, diskette, tape, or any other medium for any purpose other than backup or archival purposes.

The automobile dataset appearing on the accompanying media is Copyright © 1979 by Consumers Union of U.S., Inc., Yonkers, NY 10703-1057 and is reproduced by permission from CONSUMER REPORTS, April 1979.

Stata and Mata are registered trademarks and NetCourse is a trademark of StataCorp LP.

Other brand and product names are registered trademarks or trademarks of their respective companies.

For copyright information about the software, type `help copyright` within Stata.

The suggested citation for this software is

StataCorp. 2009. *Stata: Release 11*. Statistical Software. College Station, TX: StataCorp LP.

Table of contents

intro substantive	Introduction to multiple-imputation analysis	1
intro	Introduction to mi	14
estimation	Estimation commands for use with mi estimate	21
mi add	Add imputations from another mi dataset	24
mi append	Append mi data	27
mi convert	Change style of mi data	30
mi copy	Copy mi flongsep data	33
mi describe	Describe mi data	35
mi erase	Erase mi datasets	39
mi estimate	Estimation using multiple imputations	40
mi estimate using	Estimation using previously saved estimation results	59
mi estimate postestimation	Postestimation tools for mi estimate	66
mi expand	Expand mi data	74
mi export	Export mi data	75
mi export ice	Export mi data to ice format	76
mi export nhanes1	Export mi data to NHANES format	78
mi extract	Extract original or imputed data from mi data	81
mi import	Import data into mi	83
mi import flong	Import flong-like data into mi	86
mi import flongsep	Import flongsep-like data into mi	89
mi import ice	Import ice-format data into mi	93
mi import nhanes1	Import NHANES-format data into mi	97
mi import wide	Import wide-like data into mi	102
mi impute	Impute missing values	105
mi impute logit	Impute using logistic regression	119
mi impute mlogit	Impute using multinomial logistic regression	124
mi impute monotone	Impute missing values in monotone data	129
mi impute mvn	Impute using multivariate normal regression	145
mi impute ologit	Impute using ordered logistic regression	171
mi impute pmm	Impute using predictive mean matching	176
mi impute regress	Impute using linear regression	181
mi merge	Merge mi data	186
mi misstable	Tabulate pattern of missing values	190
mi passive	Generate/replace and register passive variables	192
mi ptrace	Load parameter-trace file into Stata	197
mi rename	Rename variable	200
mi replace0	Replace original data	203
mi reset	Reset imputed or passive variables	205
mi reshape	Reshape mi data	207
mi select	Programmer's alternative to mi extract	209
mi set	Declare multiple-imputation data	211
mi stsplit	Stsplit and stjoin mi data	215
mi update	Ensure that mi data are consistent	217
mi varying	Identify variables that vary across imputations	220
mi xeq	Execute command(s) on individual imputations	223
mi XXXset	Declare mi data to be svy, st, ts, xt, etc.	226
noupdate option	The noupdate option	228

styles	Dataset styles	230
technical	Details for programmers	238
workflow	Suggested workflow	250
Glossary		260
Subject and author index		265

Cross-referencing the documentation

When reading this manual, you will find references to other Stata manuals. For example,

[U] **26 Overview of Stata estimation commands**
[R] **regress**
[XT] **xtreg**

The first example is a reference to chapter 26, *Overview of Stata estimation commands*, in the *User's Guide*; the second is a reference to the `regress` entry in the *Base Reference Manual*; and the third is a reference to the `xtreg` entry in the *Longitudinal-Data/Panel-Data Reference Manual*.

All the manuals in the Stata Documentation have a shorthand notation:

[GSM]	*Getting Started with Stata for Mac*
[GSU]	*Getting Started with Stata for Unix*
[GSW]	*Getting Started with Stata for Windows*
[U]	*Stata User's Guide*
[R]	*Stata Base Reference Manual*
[D]	*Stata Data-Management Reference Manual*
[G]	*Stata Graphics Reference Manual*
[XT]	*Stata Longitudinal-Data/Panel-Data Reference Manual*
[MI]	*Stata Multiple-Imputation Reference Manual*
[MV]	*Stata Multivariate Statistics Reference Manual*
[P]	*Stata Programming Reference Manual*
[SVY]	*Stata Survey Data Reference Manual*
[ST]	*Stata Survival Analysis and Epidemiological Tables Reference Manual*
[TS]	*Stata Time-Series Reference Manual*
[I]	*Stata Quick Reference and Index*
[M]	*Mata Reference Manual*

Detailed information about each of these manuals may be found online at

http://www.stata-press.com/manuals/

Title

> **intro substantive** — Introduction to multiple-imputation analysis

Description

Missing data arise frequently. Various procedures have been suggested in the literature over the last several decades to deal with missing data (for example, Anderson [1957]; Hartley and Hocking [1971]; Rubin [1972, 1987]; and Dempster, Laird, and Rubin [1977]). The technique of multiple imputation, which originated in early 1970 in application to survey nonresponse (Rubin 1976), has gained popularity increasingly over the years as indicated by literature (for example, Rubin [1976, 1987, 1996]; Little [1992]; Meng [1994]; Schafer [1997]; van Buuren, Boshuizen, and Knook [1999]; Little and Rubin [2002]; Carlin et al. [2003]; Royston [2004, 2005a, 2005b, 2007]; Reiter and Raghunathan [2007]; Carlin, Galati, and Royston [2008]; and Royston, Carlin, and White [2009]).

This entry presents a general introduction to multiple imputation and describes relevant statistical terminology to be used throughout the manual. The discussion here, as well as other statistical entries in this manual, is based on the concepts developed in Rubin (1987) and Schafer (1997).

Remarks

Remarks are presented under the following headings:

> Motivating example
> What is multiple imputation?
> Theory underlying multiple imputation
> How large should M be?
> Assumptions about missing data
> Patterns of missing data
> Proper imputation methods
> Analysis of multiply imputed data
> A brief introduction to MI using Stata
> Summary

We will use the following definitions and notation.

An *imputation* represents one set of plausible values for missing data, and so *multiple imputations* represent multiple sets of plausible values. With a slight abuse of the terminology, we will use the term *imputation* to mean the data where missing values are replaced with one set of plausible values.

We use M to refer to the number of imputations and m to refer to each individual imputation, i.e., $m = 1$ means the first imputation, $m = 2$ means the second imputation, and so on.

Motivating example

Consider a fictional case–control study examining a relationship between smoking and heart attacks.

```
. use http://www.stata-press.com/data/r11/mheart0
(Fictional heart attack data; bmi missing)
. describe
Contains data from http://www.stata-press.com/data/r11/mheart0.dta
  obs:           154                          Fictional heart attack data; bmi
                                                missing
 vars:             9                          19 Jun 2009 10:50
 size:         2,926 (99.9% of memory free)

              storage  display    value
variable name   type   format     label      variable label

attack          byte   %9.0g                 Outcome (heart attack)
smokes          byte   %9.0g                 Current smoker
age             float  %9.0g                 Age, in years
bmi             float  %9.0g                 Body Mass Index, kg/m^2
female          byte   %9.0g                 Gender
hsgrad          byte   %9.0g                 High school graduate
marstatus       byte   %9.0g      mar        Marital status: single, married,
                                               divorced
alcohol         byte   %24.0g     alc        Alcohol consumption: none, <2
                                               drinks/day, >=2 drinks/day
hightar         byte   %9.0g                 Smokes high tar cigarettes

Sorted by:
```

In addition to the primary variables attack and smokes, the dataset contains information about subjects' ages, body mass indexes (BMIs), genders, educational statuses, marital statuses, alcohol consumptions, and the types of cigarettes smoked (low/high tar).

We will use logistic regression to study the relationship between attack, recording heart attacks, and smokes:

```
. logit attack smokes age bmi hsgrad female
Iteration 0:   log likelihood = -91.359017
Iteration 1:   log likelihood = -79.374749
Iteration 2:   log likelihood = -79.342218
Iteration 3:   log likelihood =  -79.34221

Logistic regression                               Number of obs   =        132
                                                  LR chi2(5)      =      24.03
                                                  Prob > chi2     =     0.0002
Log likelihood =  -79.34221                       Pseudo R2       =     0.1315

      attack |      Coef.   Std. Err.      z    P>|z|     [95% Conf. Interval]
-------------+----------------------------------------------------------------
      smokes |   1.544053   .3998329     3.86   0.000     .7603945    2.327711
         age |   .026112    .017042      1.53   0.125    -.0072898    .0595137
         bmi |   .1129938   .0500061     2.26   0.024     .0149837     .211004
      hsgrad |   .4048251   .4446019     0.91   0.363    -.4665786    1.276229
      female |   .2255301   .4527558     0.50   0.618    -.6618549    1.112915
       _cons |  -5.408398   1.810603    -2.99   0.003    -8.957115    -1.85968
```

The above analysis used 132 observations out of the available 154 because some of the covariates contain missing values. Let's examine the data for missing values, something we could have done first:

```
. misstable summarize
                                        Obs<.
                                  Unique
    Variable   Obs=.   Obs>.  Obs<.  values      Min       Max
         bmi      22            132     132  17.22643  38.24214
```

We discover that bmi is missing in 22 observations. Our analysis ignored the information about the other covariates in these 22 observations. Can we somehow preserve this information in the analysis? The answer is yes, and one solution is to use multiple imputation.

What is multiple imputation?

Multiple imputation (MI) is a flexible, simulation-based statistical technique for handling missing data. Multiple imputation consists of three steps:

1. *Imputation step.* M imputations (completed datasets) are generated under some chosen *imputation* model.

2. *Completed-data analysis (estimation) step.* The desired analysis is performed separately on each imputation $m = 1, \ldots, M$. This is called *completed-data analysis* and is the primary analysis to be performed once missing data have been imputed.

3. *Pooling step.* The results obtained from M completed-data analyses are combined into a single multiple-imputation result.

The completed-data analysis step and the pooling step can be combined and thought of generally as the analysis step.

MI as a missing-data technique has two appealing main features: 1) the ability to perform a wide variety of completed-data analyses using existing statistical methods; and 2) separation of the imputation step from the analysis step. We discuss these two features in more detail in what follows.

Among other commonly-used missing-data techniques that allow a variety of completed-data analyses are complete-case analysis or listwise (casewise) deletion, available-case analysis, and single-imputation methods. Although these procedures share one of MI's appealing properties, they lack some of MI's statistical properties.

For example, listwise deletion discards all observations with missing values and, thus, all information contained in the nonmissing values of these observations. With a large number of missing observations this may lead to results that will be less efficient (larger standard errors, wider confidence intervals, less power) than MI results. In situations when the remaining complete cases are not representative of the population of interest, listwise deletion may also lead to biased parameter estimates.

In our opening logistic analysis of heart attacks, we used listwise deletion. The effect of age was not statistically significant based on the reduced sample. The MI analysis of these data (see *A brief introduction to MI using Stata* below) will reveal the statistical significance of age by using all available observations after imputing missing values for BMI.

Unlike listwise deletion, single-imputation methods do not discard missing values. They treat the imputed values as known in the analysis. This underestimates the variance of the estimates and so overstates precision and results in confidence intervals and significance tests that are too optimistic. MI rectifies this problem by creating multiple imputations and taking into account the sampling variability due to the missing data (between-imputation variability). See Little and Rubin (2002) and Allison (2001), among others, for a more detailed comparison of the methods.

The independence of the imputation step from the analysis step is the property MI shares with other imputation methods. The imputation step fills in missing values. The analysis step provides inference about multiply imputed results and does not require any information about the missing-data aspect of the problem.

The separation of the two steps allows different individuals, a data collector/imputer and a data analyst, to perform these steps independently of one another. The advantage is that the data collector/imputer usually has access to more information about the data than may be disclosed to the data analyst and thus can create more accurate imputations. The data analyst can use the imputed data released by the data collector in a number of different analyses. Of course, it is crucial that the imputer make the imputation model as general as possible to accommodate a wide variety of analyses that the data analyst might choose to perform; see *Proper imputation methods* below for details.

In our heart attack example, the imputer would create multiple imputations of missing values of BMI using, for example, a linear regression method, and then release the resulting data to the analyst. The analyst could then analyze these multiply imputed data using an ordinary logistic regression. That is, no adjustment is needed to the analysis model itself to account for missing BMI—the pooling portion of the analysis will account for the increased variability because of imputed missing data.

Theory underlying multiple imputation

MI was derived using the Bayesian paradigm yet was proved to be statistically valid from the frequentist (randomization-based) perspective. We use the definition from Rubin (1996) of *statistical validity* that implies approximately unbiased point estimates and implies confidence intervals achieving their nominal coverages when averaged over the randomization distributions induced by the known sampling and the posited missing-data mechanisms.

To explain the role the Bayesian and frequentist concepts play in MI, we need to consider the MI procedure in more detail. MI requires specification of two models—the *imputation model* and the *analysis model*. The imputation model is the model used to create imputations in the imputation step. The analysis model is the completed-data model used during the analysis step to obtain completed-data estimates, $\widehat{Q}$, of parameters of interest, Q, and the estimate, U, of sampling variability associated with $\widehat{Q}$. During the pooling step, the individual completed-data estimates $(\widehat{Q}, U)$ are combined into $(\widehat{Q}_{\text{MI}}, T)$ to form one repeated-imputation inference. The statistical validity of the repeated-imputation inference is of interest.

Consider the case when both the imputation and analysis models are the same Bayesian models. Then the repeated imputations (multiple imputations) are repeated draws from the posterior predictive distribution of the missing data under a posited Bayesian model. The combined parameter estimates, $\widehat{Q}_{\text{MI}}$, and their associated sampling variance estimate, $T = W + B$, are the approximations to the posterior mean and variance of Q. Here W represents the within-imputation variability (average of the completed-data variance estimates, U) and B represents the between-imputation variability (variance estimate of $\widehat{Q}_{\text{MI}}$ over repeated imputations). Provided that the posterior mean and variance are adequate summaries of the posterior distribution, the repeated-imputation inference based on these combined estimates can be justified either from a purely Bayesian standpoint or from a purely frequentist standpoint. Thus a Bayesian apparatus is used to create imputations and also underlies the rules for combining parameter estimates.

In reality, the analysis model is rarely the same as the imputation model, and neither of them is an explicit Bayesian model. Repeated-imputation inference is still statistically valid in those cases. The rigorous justification is given in chapters 3 and 4 of Rubin (1987) from the frequentist perspective. Below we briefly summarize the conditions under which the repeated-imputation inference from the pooling step is statistically valid; also see Rubin (1987, 117–119) for more detail.

The repeated-imputation inference is statistically valid if 1) the multiple imputations from the imputation step are proper (see *Proper imputation methods*) and 2) the completed-data inference based on $(\widehat{Q}, U)$ from the analysis step is randomization valid. Completed-data inference based on $(\widehat{Q}, U)$ is *randomization valid* if $\widehat{Q} \sim N\{Q, \text{Var}(\widehat{Q})\}$ and U is a consistent estimate of $\text{Var}(\widehat{Q})$ over the distribution of the sampling mechanism.

The randomization validity of MI was derived under the assumption of an infinite number of imputations. In practice, however, the number of imputations tends to be small and so the finite-M properties of the MI estimators must be explored. Rubin (1987) derives the fundamental result underlying the MI inference based on a finite M. We restate it below for a scalar Q:

$$T_M^{-1/2}(Q - \widehat{Q}_M) \sim t_{\nu_M}$$

where $T_M = W + (1 + 1/M)B$ and t_{ν_M} is a Student's t distribution with degrees of freedom ν_M that depend on the number of imputations and rates of missing information (or the fraction of information missing due to nonresponse that measures the influence of the missing data on parameter estimates). Later, Li, Raghunathan, and Rubin (1991b) derived an improved procedure for multiple testing, and Barnard and Rubin (1999) and Reiter (2007) extended the MI inference to account for small samples. For computation details, see *Methods and formulas* in [MI] **mi estimate**.

How large should M be?

The theory underlying the validity of MI relies on an infinite number of imputations, M. The procedure is also known to have good statistical properties with finite M, but what values of M should we use in practice? Rubin (1987, 114) answers this question: the asymptotic relative efficiency (RE) of the MI procedure with finite M compared with infinite M is roughly 90% with only two imputations for a missing-information rate as high as 50%.

Most literature (e.g., Rubin [1987] and van Buuren, Boshuizen, and Knook [1999]) suggests that $M = 5$ (corresponding to RE of 95% for 50% of information missing) should be sufficient to obtain valid inference. In general, however, the actual number of imputations necessary for MI to perform satisfactorily depends not only on the amount of information missing due to nonresponse but also on the analysis model and the data. Some analyses may require M to be 50 or more to obtain stable results (Kenward and Carpenter 2007; Horton and Lipsitz 2001).

Literature with formal recommendations on how to choose M is very sparse. Royston (2004) and, recently, Royston, Carlin, and White (2009) discuss the impact of the number of imputations on the precision of estimates and suggest ways of determining the required number of imputations by evaluating the sampling error of the MI estimates.

Because it is computationally feasible to obtain more imputations, we recommend using at least 20 imputations to reduce the sampling error due to imputations.

Assumptions about missing data

The theory underlying MI methodology makes no assumption about the missing-data mechanism. However, many imputation methods (including those provided by Stata) require that the missing-data mechanism be ignorable. Before we discuss the ignorability conditions, consider the following definitions.

Missing data are said to be *missing completely at random* (MCAR) if the probability that data are missing does not depend on observed or unobserved data. Under MCAR, the missing data values are a simple random sample of all data values, and so any analysis that discards the missing values remains consistent, albeit perhaps inefficient.

Consider a hypothetical longitudinal study comparing different blood-pressure treatments. Suppose that the follow-up blood-pressure measurements were not collected from some subjects because they moved to a different area. These missing blood-pressure measurements can be viewed as missing completely at random as long as subjects' decisions to move were unrelated to any item in the study.

Missing data are said to be *missing at random* (MAR) if the probability that data are missing does not depend on unobserved data but may depend on observed data. Under MAR, the missing-data values do not contain any additional information given observed data about the missing-data mechanism. Note that MCAR can be viewed as a particular case of MAR. When missing data are MAR, listwise deletion may lead to biased results.

Suppose that some subjects decided to leave the study because of severe side effects from the assigned treatment of a high dosage of a medicine. Here it is unlikely that missing blood-pressure measurements are MCAR because the subjects who received a higher dosage of the medicine are more likely to suffer severe side effects than those who received a lower dosage and thus are more likely to drop out of the study. Missing blood-pressure measurements depend on the dosage of the received treatment and therefore are MAR.

On the other hand, if the subjects are withdrawn from the study for ethical reasons because of extremely high blood pressures, missing blood-pressure measurements would not be MAR. The measurements for the subjects with very high blood pressures will be missing and thus the reason for drop out will depend on the missing blood pressures. This type of missing-data mechanism is called *missing not at random* (MNAR). For such missing data, the reasons for its missingness must be accounted for in the model to obtain valid results.

Model parameters are said to be *distinct* from a Bayesian standpoint if their joint prior distribution can be factorized into independent marginal prior distributions.

The missing-data mechanism is said to be *ignorable* if missing data are MAR and the parameters of the data model and the parameters of the missing-data mechanism are distinct (Rubin 1976).

The ignorability assumption makes it possible to ignore the process that causes missing data in the imputation model, something not possible with MNAR, which simplifies the imputation step while still ensuring correct inference. The provided imputation methods assume that missing data are MAR.

In practice, it is difficult to test the ignorability assumption formally because the MAR mechanism can be distinguished from the MNAR mechanism only through the missing data that are not observed. Thus careful consideration is necessary before accepting this assumption. If in doubt, *sensitivity analysis*—analysis repeated under various missing-data models—needs to be performed to verify the stability of inference. In the context of MI, sensitivity analysis can be performed by modifying the imputation step to accommodate the nonignorable missing-data mechanism (for example, Kenward and Carpenter [2007] and van Buuren, Boshuizen, and Knook [1999]).

Patterns of missing data

Another issue we need to consider related to missing data is a *pattern of missingness* (or missing-data pattern).

Consider an $N \times p$ data matrix $Y = (Y_1, Y_2, \ldots, Y_p)'$ with p variables and N observations. Consider a permutation of column indices $(i_1, i_2, \ldots, i_p)$ such that Y_{i_1} is at least as observed as Y_{i_2}, which is at least as observed as Y_{i_3}, and so on. In other words, Y_{i_2} has missing values in the same observations (and possibly more) as Y_{i_1}, Y_{i_3} has missing values (and possibly more) in the same observations as Y_{i_2}, and so on. If such a permutation exists, then the pattern of missingness in Y is said to be *monotone*. If the pattern of missingness is not monotone, it is assumed to be arbitrary.

For example, consider the following indicator matrix recording the missing pattern in Y:

$$R_1 = \begin{pmatrix} 1 & 1 & 1 \\ 0 & 0 & 1 \\ 0 & 1 & 1 \\ 0 & 1 & 1 \end{pmatrix}$$

where R_1^{ij} is 1 if variable Y_j is observed (complete) in observation i and 0 otherwise. We can see that Y has a monotone-missing pattern if we interchange the first and the third columns of R_1. In fact, if we also rearrange the rows such that

$$R_1 = \begin{pmatrix} 1 & 1 & 1 \\ 1 & 1 & 0 \\ 1 & 1 & 0 \\ 1 & 0 & 0 \end{pmatrix}$$

then the monotonicity of missing values becomes even more evident. An example of a nonmonotone missing-value pattern is

$$R_2 = \begin{pmatrix} 1 & 1 & 1 \\ 1 & 1 & 0 \\ 0 & 1 & 0 \\ 1 & 0 & 0 \end{pmatrix}$$

There is no ordering of the first two columns of R_2 such that the missing values in one column imply missing values in the other column.

Why is it important to consider the monotone missing-value pattern? A monotone-missing pattern greatly simplifies the imputation task. Under a monotone-missing pattern, a multivariate imputation task can be formulated as a sequence of independent univariate (conditional) imputation tasks, which allows the creation of a flexible imputation model; see [MI] **mi impute monotone** for details, and see Rubin (1987, 174) for more technical conditions under which such a formulation is applicable.

Proper imputation methods

As we mentioned earlier, a key concept underlying the randomization-based evaluations of the repeated-imputation inference is proper multiple imputation.

A multiple-imputation method is said to be *proper* if it produces proper multiple imputations, which we are about to define. Rubin (1987, 118–119) gives a full technical definition for proper multiple imputations. Ignoring the more technical definition, Rubin (1996) states the following main conditions. The multiple imputations are said to be *proper* if

1. MI estimates $\widehat{Q}_{\mathrm{MI}}$ are asymptotically normal with mean $\widehat{Q}$ and a consistent variance–covariance estimate B.

2. The within-imputation variance estimate W is a consistent estimate of the variance–covariance estimate U with variability of a lower order than $\mathrm{Var}(\widehat{Q}_{\mathrm{MI}})$.

The above statements assume a large number of imputations and the randomization distribution induced by the missing-data mechanism.

In general, it is difficult to determine if an imputation method is proper using the above definition. Rubin (1987, sec. 4.3) and Binder and Sun (1996) describe several examples of proper and improper imputation methods. Rubin (1987, 125–127) recommends drawing imputations from a Bayesian posterior predictive distribution (or an appropriate approximation to it) of missing values under the chosen model for the data and the missing-data mechanism. The chosen imputation model must also be appropriate for the completed-data statistics likely to be used at the analysis stage. Schafer (1997, 145) points out that from a practical standpoint, it is more important that the chosen imputation model performs well over the repeated samples than that it be technically proper. This can be checked via simulation.

With the exception of predictive mean matching, the imputation methods available in Stata obtain imputations by simulating from a Bayesian posterior predictive distribution of the missing data (or its approximation) under the conventional (or chosen) prior distribution; see *Imputation methods* in [MI] **mi impute** for details. To ensure that the multiple imputations are proper, you must choose an appropriate imputation model, which we briefly discuss next.

The imputation model must include all predictors relevant to the missing-data mechanism, and it must preserve all data characteristics likely to be explored at the analysis stage. For example, if the analysis model explores a correlation between two variables, then omitting either of those variables from the imputation model will lead to estimates of the correlation biased toward zero. Another common mistake that may lead to biased estimates is when an outcome variable of the analysis model is not used in the imputation model. In the survey context, all structural variables such as sampling weights, strata, and cluster identifiers (or at least main strata and main clusters) need to be included in the imputation model.

In general, any predictors involved in the definition of the completed-data estimators and the sampling design should be included in the imputation model. If you intend to use the multiply imputed data in an analysis involving a wide range of completed-data estimators, you should include as many variables as possible.

Using our heart attack data, if we were to release the multiply imputed version of it for general analyses, we would have included all available covariates as predictors in the regression model used to impute BMI and not only the subset of covariates (heart attacks, smoking status, age, gender, and educational status) used in our specific data analysis.

The severity of the effect of a misspecified imputation model will typically depend on the amount of imputed data relative to the observed data—a small number of observations with improperly imputed values may not affect the inference greatly if there is a large number of observations with complete data.

For more details about imputation modeling, see Rubin (1996), Schafer (1997, 139–144), Schafer and Olsen (1998), Allison (2001), Schafer and Graham (2002), Kenward and Carpenter (2007), and Graham (2009), among others. For imputation modeling of large surveys, see, for example, Schafer, Khare, and Ezzati-Rice (1993) and Ezzati-Rice et al. (1995).

Analysis of multiply imputed data

Once we have multiply imputed data, we perform our primary analysis on each completed dataset and then use Rubin's combination rules to form one set of results. Assuming that the underlying imputation model is properly specified, we can choose from a variety of statistical methods. For example, the methods can include maximum-likelihood methods, survey methods, nonparametric methods, and any other method appropriate for the type of data we have.

Each of the methods have certain concepts associated with them. For example, maximum-likelihood methods use a likelihood function, whereas a deviance is associated with generalized linear models. While these concepts are well defined within each individual completed-data analysis, they may not have a clear interpretation when the individual analyses are combined in the pooling step. (Only in the special case when the imputation and analysis models are compatible Bayesian models can the estimated parameters be viewed as approximations to the mode of the posterior distribution.)

As a result, various statistical (postestimation) procedures based on these concepts, such as likelihood-ratio tests, goodness-of-fit tests, etc., are not directly applicable to MI results. Instead, their "MI" versions are being studied in the literature (Li et al. 1991a; Meng and Rubin 1992). Another concept that is not well defined within MI is that of prediction, although some insight into it has been given by Carlin, Galati, and Royston (2008).

Donald Bruce Rubin (1943–) was born in Washington, DC. He entered Princeton intending to become a physicist, but ended up majoring in psychology. He entered Harvard intending to continue as a psychologist, but in the event, gained further degrees in computer science and statistics. After periods at the Educational Testing Service and the University of Chicago, Rubin returned to Harvard in 1984. He has had many visiting appointments and has carried out extensive consultancy work. Rubin has long been a leader in research on causal inference in experiments and observational studies, and problems of nonresponse and missing data. Among many major contributions is his formalization of the expectation-maximization algorithm with Arthur Dempster and Nan Laird. Rubin's work ranges over a wide variety of sciences and is often Bayesian in style.

A brief introduction to MI using Stata

Stata offers full support for MI analysis from the imputation step to the pooling step.

The imputation step can be performed for one variable or multiple variables. A number of imputation methods, including flexible methods accommodating variables of different types for monotone-missing data and an iterative MCMC method based on MVN for arbitrary missingness, are available; see [MI] **mi impute** for details.

The analysis and pooling steps are combined into one step and performed by `mi estimate`; see [MI] **mi estimate**. You can fit many commonly used models and obtain combined estimates of coefficients (or transformed coefficients) (see [MI] **estimation** for a list of supported estimation commands), or you can create your own estimation command and use it with the `mi estimate` prefix.

In addition to the conventional estimation steps, Stata facilitates many data-manipulation routines for managing your multiply imputed data and verifying its integrity over the imputations; see [MI] **intro** for a full list of commands.

As a short demonstration of `mi`, let's analyze the heart attack data introduced earlier using MI; see [MI] **workflow** for more thorough guidelines.

The goals are 1) to fill in missing values of `bmi` using, for example, a linear regression imputation method (`mi impute regress`) to obtain multiply imputed data and 2) to analyze the multiply imputed data using logistic regression, which we will do using `mi estimate`. Before we can accomplish these two steps, we need to prepare the data so they can be used with `mi`. First, we declare the data to be `mi` data:

```
. use http://www.stata-press.com/data/r11/mheart0
(Fictional heart attack data; bmi missing)
. mi set mlong
```

We choose to use the data in the marginal long style (`mlong`) because it is a memory-efficient style; see [MI] **styles** for details.

To use `mi impute`, we must first register imputation variables. In general, we recommend that you register all variables relevant to the analysis as `imputed`, `passive`, or `regular` with `mi register` (see [MI] **mi set**), especially if you plan on doing any data management of your multiply imputed data.

```
. mi register imputed bmi
(22 m=0 obs. now marked as incomplete)
. mi register regular attack smokes age hsgrad female
```

We are now ready to use `mi impute`. To lessen the simulation (Monte Carlo) error, we arbitrarily choose to create 20 imputations (`add(20)` option). We also specify the `rseed()` option for reproducibility:

```
. mi impute regress bmi attack smokes age hsgrad female, add(20) rseed(2232)
Univariate imputation               Imputations =        20
Linear regression                         added =        20
Imputed: m=1 through m=20               updated =         0
```

	Observations per m			
Variable	complete	incomplete	imputed	total
bmi	132	22	22	154

```
(complete + incomplete = total; imputed is the minimum across m
 of the number of filled in observations.)
```

From the output, we see that all 22 incomplete values of `bmi` were successfully imputed. You may want to examine your imputations to verify that nothing abnormal occurred during imputation. For example, as a quick check, we can compare main descriptive statistics from some imputations (say, the first and the last one) to those from the observed data. We use `mi xeq` (see [MI] **mi xeq**) to execute Stata's `summarize` command on the original data ($m = 0$), the first imputation ($m = 1$), and the last imputation ($m = 20$):

```
. mi xeq 0 1 20: summarize bmi

m=0 data:
-> summarize bmi
```

Variable	Obs	Mean	Std. Dev.	Min	Max
bmi	132	25.24136	4.027137	17.22643	38.24214

```
m=1 data:
-> summarize bmi
```

Variable	Obs	Mean	Std. Dev.	Min	Max
bmi	154	25.0776	4.016672	17.22643	38.24214

```
m=20 data:
-> summarize bmi
```

Variable	Obs	Mean	Std. Dev.	Min	Max
bmi	154	25.41818	4.114603	14.71254	38.24214

The summary statistics of the imputed datasets look reasonable.

We now fit the logistic regression using the `mi estimate` prefix command:

```
. mi estimate, dots: logit attack smokes age bmi hsgrad female

Imputations (20):
.........10.........20 done

Multiple-imputation estimates          Imputations       =         20
Logistic regression                    Number of obs     =        154
                                       Average RVI       =     0.0312
DF adjustment:   Large sample          DF:      min      =    1060.38
                                                avg      =  223362.56
                                                max      =  493335.88
Model F test:        Equal FMI         F(   5, 71379.3)  =       3.59
Within VCE type:          OIM          Prob > F          =     0.0030

------------------------------------------------------------------------------
      attack |      Coef.   Std. Err.      t    P>|t|     [95% Conf. Interval]
-------------+----------------------------------------------------------------
      smokes |   1.198595   .3578195     3.35   0.001     .4972789    1.899911
         age |   .0360159   .0154399     2.33   0.020     .0057541    .0662776
         bmi |   .1039416   .0476136     2.18   0.029      .010514    .1973692
      hsgrad |   .1578992   .4049257     0.39   0.697    -.6357464    .9515449
      female |  -.1067433   .4164735    -0.26   0.798    -.9230191    .7095326
       _cons |  -5.478143   1.685075    -3.25   0.001    -8.782394   -2.173892
------------------------------------------------------------------------------
```

Compared with the earlier `logit` analysis (using listwise deletion), we detect the significance of `age`, whose effect was apparently disguised by the missing data. See [MI] **mi estimate** for details.

We will be using variations of these data throughout the `mi` documentation.

Summary

- MI is a simulation-based procedure. Its purpose is not to re-create the individual missing values as close as possible to the true ones but to handle missing data in a way resulting in valid statistical inference (Rubin 1987, 1996).

- MI yields valid inference if 1) the imputation method is proper with respect to the posited missing-data mechanism (see *Proper imputation methods*), and 2) completed-data analysis is valid in the absence of missing data.

- A small number of imputations (5 to 20) may be sufficient when fractions of missing data are low. High fractions of missing data as well as particular data structures may require up to 100 (or more) imputations. Whenever feasible to do so, we recommend that you vary the number of imputations to see if this affects your results.

- With a small number of imputations, the reference distribution for the MI inference is Student's t (or F in multiple-hypothesis testing). The residual degrees of freedom depend on M and the rates of missing information and thus are different for each parameter of interest.

- With a large number of imputations, the reference distribution for MI inference is approximately normal (or χ^2 in multiple-hypothesis testing).

- When the imputer's model is more restrictive than the analyst's model, the MI inference can be invalid if the imputer's assumptions are not true. On the other hand, when the analyst's model is more restrictive than the imputer's model, the MI results will be valid but somewhat conservative if the analyst's assumptions are true. If the analyst's assumptions are false, the results can be biased; see, for example, Schafer (1997) for details.

- MI is relatively robust to departures from the correct specification of the imputation model, provided the rates of missing information are low and the correct completed-data model is used in the analysis.

- Certain concepts, e.g., likelihood and deviance, do not have clear interpretation within the MI framework. As such, various statistical (postestimation) procedures based on these concepts (e.g., likelihood-ratio tests, goodness-of-fit tests) are not directly applicable to MI results.

References

Allison, P. D. 2001. *Missing Data*. Thousand Oaks, CA: Sage.

Anderson, T. W. 1957. Maximum likelihood estimates for a multivariate normal distribution when some observations are missing. *Journal of the American Statistical Association* 52: 200–203.

Barnard, J., and D. B. Rubin. 1999. Small-sample degrees of freedom with multiple imputation. *Biometrika* 86: 948–955.

Binder, D. A., and W. Sun. 1996. Frequency valid multiple imputation for surveys with a complex design. *Proceedings of the Survey Research Methods Section, American Statistical Association* 281–286.

Carlin, J. B., J. C. Galati, and P. Royston. 2008. A new framework for managing and analyzing multiply imputed data in Stata. *Stata Journal* 8: 49–67.

Carlin, J. B., N. Li, P. Greenwood, and C. Coffey. 2003. Tools for analyzing multiple imputed datasets. *Stata Journal* 3: 226–244.

Dempster, A. P., N. M. Laird, and D. B. Rubin. 1977. Maximum likelihood from incomplete data via the EM algorithm. *Journal of the Royal Statistical Society, Series B* 39: 1–38.

Ezzati-Rice, T. M., W. Johnson, M. Khare, R. J. A. Little, D. B. Rubin, and J. L. Schafer. 1995. A simulation study to evaluate the performance of model-based multiple imputations in NCHS health examination surveys. *Proceedings of the Annual Research Conference*, 257–266. U.S. Bureau of the Census: Washington, DC.

Gelman, A., J. B. Carlin, H. S. Stern, and D. B. Rubin. 2004. *Bayesian Data Analysis*. 2nd ed. London: Chapman & Hall/CRC.

Gelman, A., and D. B. Rubin. 1992. Inference from iterative simulation using multiple sequences. *Statistical Science* 7: 457–472.

Graham, J. W. 2009. Missing data analysis: Making it work in the real world. *Annual Review of Psychology* 60: 549–576.

Hartley, H. O., and R. R. Hocking. 1971. The analysis of incomplete data (with discussion). *Biometrics* 27: 783–823.

Horton, N. J., and K. P. Kleinman. 2007. Much ado about nothing: A comparison of missing data methods and software to fit incomplete data regression models. *American Statistician* 61: 79–90.

Horton, N. J., and S. R. Lipsitz. 2001. Multiple imputation in practice: Comparison of software packages for regression models with missing variables. *American Statistician* 55: 244–254.

Kenward, M. G., and J. R. Carpenter. 2007. Multiple imputation: Current perspectives. *Statistical Methods in Medical Research* 16: 199–218.

Li, K.-H. 1988. Imputation using Markov chains. *Journal of Statistical Computation and Simulation* 30: 57–79.

Li, K.-H., X.-L. Meng, T. E. Raghunathan, and D. B. Rubin. 1991a. Significance levels from repeated p-values with multiply-imputed data. *Statistica Sinica* 1: 65–92.

Li, K.-H., T. E. Raghunathan, and D. B. Rubin. 1991b. Large-sample significance levels from multiply imputed data using moment-based statistics and an F reference distribution. *Journal of the American Statistical Association* 86: 1065–1073.

Little, R. J. A. 1988. Missing-data adjustments in large surveys. *Journal of Business and Economic Statistics* 6: 287–296.

———. 1992. Regression with missing X's: A review. *Journal of the American Statistical Association* 87: 1227–1237.

Little, R. J. A., and D. B. Rubin. 2002. *Statistical Analysis with Missing Data*. 2nd ed. Hoboken, NJ: Wiley.

Meng, X.-L. 1994. Multiple-imputation inferences with uncongenial sources of input (with discussion). *Statistical Science* 9: 538–573.

Meng, X.-L., and D. B. Rubin. 1992. Performing likelihood ratio tests with multiply-imputed data sets. *Biometrika* 79: 103–111.

Raghunathan, T. E., J. M. Lepkowski, J. Van Hoewyk, and P. Solenberger. 2001. A multivariate technique for multiply imputing missing values using a sequence of regression models. *Survey Methodology* 27: 85–95.

Reiter, J. P. 2007. Small-sample degrees of freedom for multi-component significance tests with multiple imputation for missing data. *Biometrika* 94: 502–508.

———. 2008. Multiple imputation when records used for imputation are not used or disseminated for analysis. *Biometrika* 95: 933–946.

Reiter, J. P., and T. E. Raghunathan. 2007. The multiple adaptations of multiple imputation. *Journal of the American Statistical Association* 102: 1462–1471.

Royston, P. 2004. Multiple imputation of missing values. *Stata Journal* 4: 227–241.

———. 2005a. Multiple imputation of missing values: Update. *Stata Journal* 5: 188–201.

———. 2005b. Multiple imputation of missing values: Update of ice. *Stata Journal* 5: 527–536.

———. 2007. Multiple imputation of missing values: Further update of ice, with an emphasis on interval censoring. *Stata Journal* 7: 445–464.

Royston, P., J. B. Carlin, and I. R. White. 2009. Multiple imputation of missing values: New features for mim. *Stata Journal*. Forthcoming.

Rubin, D. B. 1972. A non-iterative algorithm for least squares estimation of missing values in any analysis of variance design. *Journal of the Royal Statistical Society, Series C* 21: 136–141.

———. 1976. Inference and missing data. *Biometrika* 63: 581–592.

———. 1986. Statistical matching using file concatenation with adjusted weights and multiple imputations. *Journal of Business and Economic Statistics* 4: 87–94.

———. 1987. *Multiple Imputation for Nonresponse in Surveys*. New York: Wiley.

———. 1996. Multiple imputation after 18+ years. *Journal of the American Statistical Association* 91: 473–489.

Schafer, J. L. 1997. *Analysis of Incomplete Multivariate Data*. Boca Raton, FL: Chapman & Hall/CRC.

Schafer, J. L., and J. W. Graham. 2002. Missing data: Our view of the state of the art. *Psychological Methods* 7: 147–177.

Schafer, J. L., M. Khare, and T. M. Ezzati-Rice. 1993. Multiple imputation of missing data in NHANES III. *Proceedings of the Annual Research Conference*, 459–487. U.S. Bureau of the Census: Washington, DC.

Schafer, J. L., and M. K. Olsen. 1998. Multiple imputation for multivariate missing-data problems: A data analyst's perspective. *Multivariate Behavioral Research* 33: 545–571.

Schenker, N., and J. M. G. Taylor. 1996. Partially parametric techniques for multiple imputation. *Computational Statistics & Data Analysis* 22: 425–446.

Tanner, M. A., and W. H. Wong. 1987. The calculation of posterior distributions by data augmentation (with discussion). *Journal of the American Statistical Association* 82: 528–550.

van Buuren, S. 2007. Multiple imputation of discrete and continuous data by fully conditional specification. *Statistical Methods in Medical Research* 16: 219–242.

van Buuren, S., H. C. Boshuizen, and D. L. Knook. 1999. Multiple imputation of missing blood pressure covariates in survival analysis. *Statistics in Medicine* 18: 681–694.

Also see

[MI] **intro** — Introduction to mi

[MI] **workflow** — Suggested workflow

[MI] **mi impute** — Impute missing values

[MI] **estimation** — Estimation commands for use with mi estimate

[MI] **mi estimate** — Estimation using multiple imputations

[MI] **Glossary**

Title

> **intro** — Introduction to mi

Syntax

> To become familiar with mi as quickly as possible, do the following:
> 1. See *A simple example* under *Remarks* below.
> 2. If you have data that require imputing, see [MI] **mi set** and [MI] **mi impute**.
> 3. Alternatively, if you have already imputed data, see [MI] **mi import**.
> 4. To fit your model, see [MI] **mi estimate**.

To create mi data from original data

mi set	declare data to be mi data
mi register	register imputed, passive, or regular variables
mi unregister	unregister previously registered variables
mi unset	return data to unset status (rarely used)

See *Description* below for a summary of mi data and these commands.
See [MI] **Glossary** for a definition of terms.

To import data that already have imputations for the missing values, do not mi set the data, but instead use

mi import	import mi data
mi export	export mi data to non-Stata application

Once data are mi set or mi imported

mi query	query whether and how mi set
mi describe	describe mi data
mi varying	identify variables that vary over m
mi misstable	tabulate missing values
mi passive	create passive variable and register it

To perform estimation on mi data

mi impute	impute missing values
mi estimate	perform and combine estimation on $m > 0$
mi ptrace	check stability of MCMC
mi test	perform tests on coefficients
mi testtransform	perform tests on transformed coefficients

To stset, svyset, tsset, or xtset any mi data that were not set at the time they were mi set

mi fvset	fvset for mi data
mi svyset	svyset for mi data
mi xtset	xtset for mi data
mi tsset	tsset for mi data
mi stset	stset for mi data
mi streset	streset for mi data
mi st	st for mi data

To perform data management on mi data

mi rename	rename variable
mi append	append for mi data
mi merge	merge for mi data
mi expand	expand for mi data
mi reshape	reshape for mi data
mi stsplit	stsplit for mi data
mi stjoin	stjoin for mi data
mi add	add imputations from one mi dataset to another

To perform data management for which no mi prefix command exists

mi extract 0	extract $m = 0$ data
...	perform data management the usual way
mi replace0	replace $m = 0$ data in mi data

To perform the same data-management or data-reporting command(s) on $m = 0, m = 1, \ldots$

mi xeq: ...	execute commands on $m = 0, m = 1, m = 2, \ldots, m = M$
mi xeq #: ...	execute commands on $m = \#$
mi xeq # # ...: ...	execute commands on specified values of m

Useful utility commands include

mi convert	convert mi data from one *style* to another
mi extract #	extract $m = \#$ from mi data
mi select #	programmer's command similar to mi extract
mi copy	copy mi data
mi erase	erase files containing mi data
mi update	verify/make mi data consistent
mi reset	reset imputed or passive variable

Programmers interested in extending mi should see

[MI] **technical**	Detail for programmers

Summary of styles

There are four styles or formats in which mi data are stored: flongsep, flong, mlong, and wide.

1. Flongsep: $m = 0, m = 1, \ldots, m = M$ are each separate .dta datasets. If $m = 0$ data are stored in pat.dta, then $m = 1$ data are stored in _1_pat.dta, $m = 2$ in _2_pat.dta, and so on. Flongsep stands for *full long and separate*.

2. Flong: $m = 0, m = 1, \ldots, m = M$ are stored in one dataset with $\_N = N + M \times N$ observations, where N is the number of observations in $m = 0$. Flong stands for *full long*.

3. Mlong: $m = 0, m = 1, \ldots, m = M$ are stored in one dataset with $\_N = N + M \times n$ observations, where n is the number of incomplete observations in $m = 0$. Mlong stands for *marginal long*.

4. Wide: $m = 0, m = 1, \ldots, m = M$ are stored in one dataset with $\_N = N$ observations. Each imputed and passive variable has M additional variables associated with it. If variable bp contains the values in $m = 0$, then values for $m = 1$ are contained in variable _1_bp, values for $m = 2$ in _2_bp, and so on. Wide stands for *wide*.

See style in [MI] **Glossary** and see [MI] **styles** for examples. See [MI] **technical** for programmer's details.

Description

The mi suite of commands deals with multiple-imputation data, abbreviated as mi data.

In summary,

1. mi data may be stored in one of four formats—flongsep, flong, mlong, and wide—known as styles. Descriptions are provided in *Summary of styles* directly above.

2. mi data contain M imputations numbered $m = 1, 2, \ldots, M$, and contain $m = 0$, the original data with missing values.

3. Each variable in `mi` data is registered as imputed, passive, or regular, or it is unregistered.

   a. Unregistered variables are mostly treated like regular variables.

   b. Regular variables usually do not contain missing, or if they do, the missing values are not imputed in $m > 0$.

   c. Imputed variables contain missing in $m = 0$, and those values are imputed, or to be imputed, in $m > 0$.

   d. Passive variables are algebraic combinations of imputed, regular, or other passive variables.

4. If an imputed variable contains a value greater than . in $m = 0$—it contains .a, .b, ..., .z—then that value is considered a hard missing and the missing value persists in $m > 0$.

See [MI] **Glossary** for a more thorough description of terms used throughout this manual.

All `mi` commands are implemented as ado-files.

Remarks

Remarks are presented under the following headings:

> *A simple example*
> *Suggested reading order*

A simple example

We are about to type six commands:

```
. use http://www.stata-press.com/data/r11/mheart5                (1)
. mi set mlong                                                   (2)
. mi register imputed age bmi                                    (3)
. set seed 29390                                                 (4)
. mi impute mvn age bmi = attack smokes hsgrad female, add(10)   (5)
. mi estimate: logistic attack smokes age bmi hsgrad female      (6)
```

The story is that we want to fit

```
. logistic attack smokes age bmi hsgrad female
```

but the `age` and `bmi` variables contain missing values. Fitting the model by typing `logistic` ... would ignore some of the information in our data. Multiple imputation (MI) attempts to recover that information. The method imputes M values to fill in each of the missing values. After that, statistics are performed on the M imputed datasets separately and the results combined. The goal is to obtain better estimates of parameters and their standard errors.

In the solution shown above,

1. We load the data.

2. We set our data for use with `mi`.

3. We inform `mi` which variables contain missing values for which we want to impute values.

4. We impute values in command 5; we prefer that our results be reproducible, so we set the random-number seed in command 4. This step is optional.

5. We create $M = 10$ imputations for each missing value in the variables we registered in command 3.

6. We fit the desired model separately on each of the 10 imputed datasets and combine the results.

The results of running the six-command solution are

```
. use http://www.stata-press.com/data/r11/mheart5
(Fictional heart attack data, bmi and age missing)
. mi set mlong
. mi register imputed age bmi
(28 m=0 obs. now marked as incomplete)
. set seed 29390
. mi impute mvn age bmi = attack smokes hsgrad female, add(10)
Performing EM optimization:
note: 12 observations omitted from EM estimation because of all imputation
      variables missing
  observed log likelihood = -651.75868 at iteration 7
Performing MCMC data augmentation ...
```

```
Multivariate imputation              Imputations  =         10
Multivariate normal regression             added  =         10
Imputed: m=1 through m=10                updated  =          0

Prior: uniform                        Iterations  =       1000
                                         burn-in  =        100
                                         between  =        100
```

	Observations per m			
Variable	complete	incomplete	imputed	total
age	142	12	12	154
bmi	126	28	28	154

(complete + incomplete = total; imputed is the minimum across m
 of the number of filled in observations.)

```
. mi estimate: logistic attack smokes age bmi hsgrad female
Multiple-imputation estimates              Imputations    =         10
Logistic regression                        Number of obs  =        154
                                           Average RVI    =     0.0682
DF adjustment:     Large sample            DF:     min    =     243.02
                                                   avg    =   12704.39
                                                   max    =   31046.43
Model F test:       Equal FMI              F(   5, 7017.5) =       3.50
Within VCE type:         OIM               Prob > F       =     0.0037
```

attack	Coef.	Std. Err.	t	P>|t|	[95% Conf. Interval]	
smokes	1.233068	.3642825	3.38	0.001	.5190339	1.947102
age	.0336572	.0165166	2.04	0.042	.0012368	.0660777
bmi	.1136643	.0498647	2.28	0.024	.0154421	.2118866
hsgrad	.1795391	.4078177	0.44	0.660	-.6198001	.9788783
female	-.0936591	.4193291	-0.22	0.823	-.9155651	.7282469
_cons	-5.613589	1.768236	-3.17	0.002	-9.092009	-2.135169

Note that the output from the last command,

. mi estimate: logistic attack smokes age bmi hsgrad female

reported coefficients rather than odds ratios, which logistic would usually report. That is because the estimation command is not logistic, it is mi estimate, and mi estimate happened to use logistic to obtain results that mi estimate combined into its own estimation results.

Anyway, `mi estimate` by default displays coefficients. If we now wanted to see odds ratios, we could type

    `. mi estimate, or`
    (*output showing odds ratios would appear*)

Note carefully: We replay results by typing `mi estimate`, not by typing `logistic`. If we had wanted to see the odds ratios from the outset, we would have typed

    `. mi estimate, or: logistic attack smokes age bmi hsgrad female`

Suggested reading order

The order of suggested reading of this manual is

    [MI] **intro substantive**
    [MI] **intro**
    [MI] **Glossary**
    [MI] **workflow**

    [MI] **mi set**
    [MI] **mi import**
    [MI] **mi describe**
    [MI] **mi misstable**

    [MI] **mi impute**
    [MI] **mi estimate**

    [MI] **styles**
    [MI] **mi convert**
    [MI] **mi update**

    [MI] **mi rename**
    [MI] **mi copy**
    [MI] **mi erase**
    [MI] **mi XXXset**

    [MI] **mi extract**
    [MI] **mi replace0**

    [MI] **mi append**
    [MI] **mi add**
    [MI] **mi merge**
    [MI] **mi reshape**
    [MI] **mi stsplit**
    [MI] **mi varying**

Programmers will want to see [MI] **technical**.

Acknowledgments

We thank Jerry (Jerome) Reiter of Duke University, Patrick Royston of the MRC Clinical Trials Unit, and Ian White of the MRC Biostatistics Unit for their comments and assistance in the development of mi. We also thank James Carpenter of the London School of Hygiene and Tropical Medicine and Jonathan Sterne of the University of Bristol for their comments.

Previous and still ongoing work on multiple imputation in Stata influenced the design of mi. For their past and current contributions, we thank Patrick Royston and Ian White again for ice; John Carlin and John Galati, both of the Murdoch Children's Research Institute and University of Melbourne, and Patrick Royston and Ian White (yet again) for mim; John Galati for inorm; and Rodrigo Alfaro of the Banco Central de Chile for mira.

Also see

[MI] **intro substantive** — Introduction to multiple-imputation analysis

[MI] **Glossary**

[MI] **styles** — Dataset styles

[MI] **workflow** — Suggested workflow

Title

estimation — Estimation commands for use with mi estimate

Description

Multiple-imputation data analysis in Stata is similar to the standard data analysis. The standard syntax applies, but you need to remember the following for MI data analysis:

1. The data must be declared as mi data.

    If you already have multiply imputed data (saved in Stata format), use mi import to import it into mi; see [MI] **mi import**.

    If you do not have multiply imputed data, use mi set (see [MI] **mi set**) to declare your original data to be mi data and use mi impute (see [MI] **mi impute**) to fill in missing values.

2. Prefix the estimation commands with mi estimate: (see [MI] **mi estimate**).

(*Continued on next page*)

estimation — Estimation commands for use with mi estimate

The following estimation commands support the `mi estimate` prefix.

command	entry	description
Linear regression models		
regress	[R] **regress**	Linear regression
cnsreg	[R] **cnsreg**	Constrained linear regression
mvreg	[R] **mvreg**	Multivariate regression
Binary-response regression models		
logistic	[R] **logistic**	Logistic regression, reporting odds ratios
logit	[R] **logit**	Logistic regression, reporting coefficients
probit	[R] **probit**	Probit regression
cloglog	[R] **cloglog**	Complementary log-log regression
binreg	[R] **binreg**	GLM for the binomial family
Count-response regression models		
poisson	[R] **poisson**	Poisson regression
nbreg	[R] **nbreg**	Negative binomial regression
gnbreg	[R] **nbreg**	Generalized negative binomial regression
Ordinal-response regression models		
ologit	[R] **ologit**	Ordered logistic regression
oprobit	[R] **oprobit**	Ordered probit regression
Categorical-response regression models		
mlogit	[R] **mlogit**	Multinomial (polytomous) logistic regression
mprobit	[R] **mprobit**	Multinomial probit regression
clogit	[R] **clogit**	Conditional (fixed-effects) logistic regression
Quantile regression models		
qreg	[R] **qreg**	Quantile regression
iqreg	[R] **qreg**	Interquantile range regression
sqreg	[R] **qreg**	Simultaneous-quantile regression
bsqreg	[R] **qreg**	Quantile regression with bootstrap standard errors
Survival regression models		
stcox	[ST] **stcox**	Cox proportional hazards model
streg	[ST] **streg**	Parametric survival models
stcrreg	[ST] **stcrreg**	Competing-risks regression
Other regression models		
glm	[R] **glm**	Generalized linear models
areg	[R] **areg**	Linear regression with a large dummy-variable set
rreg	[R] **rreg**	Robust regression
truncreg	[R] **truncreg**	Truncated regression
Descriptive statistics		
mean	[R] **mean**	Estimate means
proportion	[R] **proportion**	Estimate proportions
ratio	[R] **ratio**	Estimate ratios
Survey regression models		
svy:	[SVY] **svy**	Estimation commands for survey data (excluding commands that are not listed above)

Also see

[MI] **mi estimate** — Estimation using multiple imputations

[MI] **mi estimate postestimation** — Postestimation tools for mi estimate

[MI] **mi set** — Declare multiple-imputation data

[MI] **mi import** — Import data into mi

[MI] **mi impute** — Impute missing values

[MI] **workflow** — Suggested workflow

[MI] **intro substantive** — Introduction to multiple-imputation analysis

[MI] **intro** — Introduction to mi

[MI] **Glossary**

Title

mi add — Add imputations from another mi dataset

Syntax

mi add *varlist* using *filename* [, *options*]

options	description
assert(master)	assert all observations found in master
assert(match)	assert all observations found in master and in using
noupdate	see [MI] **noupdate option**

Notes:

1. Jargon:
   match variables = *varlist*, variables on which match performed
   master = data in memory
   using = data on disk (*filename*)
2. Master must be mi set.
3. Using must be mi set.
4. *filename* must be enclosed in double quotes if *filename* contains blanks or other special characters.

Menu

Statistics > Multiple imputation

Description

mi add adds the imputations from the using to the end of the master.

Options

assert(*results*) specifies how observations are expected to match. If results are not as you expect, an error message will be issued and the master data left unchanged.

assert(master) specifies that you expect a match for every observation in the master, although there may be extra observations in the using that mi add is to ignore.

assert(match) specifies that you expect every observation in the master to match an observation in the using and vice versa.

The default is that the master may have observations that are missing from the using and vice versa. Only observations in common are used by mi add.

noupdate in some cases suppresses the automatic mi update this command might perform; see [MI] **noupdate option**.

Remarks

Think of the result produced by `mi add` as being

Result	Source
$m = 0$	$m = 0$ from master
$m = 1$	$m = 1$ from master
$m = 2$	$m = 2$ from master
.	.
.	.
.	.
$m = M_{\text{master}}$	$m = M_{\text{master}}$ from master
$m = M_{\text{master}} + 1$	$m = 1$ from using
$m = M_{\text{master}} + 2$	$m = 2$ from using
.	.
.	.
.	.
$m = M_{\text{master}} + M_{\text{using}}$	$m = M_{\text{using}}$ from using

That is, the original data in the master remain unchanged. All that happens is the imputed data from the using are added to the end of the master as additional imputations.

For instance, say you discover that you and a coworker have been working on the same data. You have added $M = 20$ imputations to your data. Your coworker has separately added $M = 17$. To combine the data, type something like

```
. use mydata
. mi add patientid using karensdata
(17 imputations added; M=37)
```

The only thing changed in your data is M. If your coworker's data have additional variables, they are ignored. If your coworker has variables registered differently from how you have them registered, that is ignored. If your coworker has not yet registered as imputed a variable that you have registered as imputed, that is noted in the output. You might see

```
. use mydata
. mi add patientid using karensdata
(17 imputations added; M=37)
(imputed variable grade not found in using data;
    added imputations contain m=0 values for that variable)
```

Saved results

`mi add` saves the following in `r()`:

Scalars
    `r(m)`   number of added imputations
    `r(unmatched_m)`   number of unmatched master observations
    `r(unmatched_u)`   number of unmatched using observations

Macros
    `r(imputed_f)`   variables for which imputed found
    `r(imputed_nf)`   variables for which imputed not found

Also see

[MI] **intro** — Introduction to mi

[MI] **mi append** — Append mi data

[MI] **mi merge** — Merge mi data

Title

mi append — Append mi data

Syntax

mi append using *filename* [, *options*]

options	description
<u>g</u>enerate(*newvar*)	create *newvar*; 0 = master, 1 = using
<u>nol</u>abel	do not copy value labels from using
<u>non</u>otes	do not copy notes from using
force	string $\leftrightarrow$ numeric not type mismatch error
noupdate	see [MI] **noupdate option**

Notes:

1. Jargon:
   master = data in memory
   using = data on disk (*filename*.dta)

2. Master must be mi set; using may be mi set.

3. mi append is syntactically and logically equivalent to append; see [D] **append**. The resulting data have $M = \max(M_{\text{master}}, M_{\text{using}})$, not their sum. See [MI] **mi add** to append imputations holding $m = 0$ constant.

4. mi append syntactically differs from append in that multiple using files may not be specified and the keep(*varlist*) option is not allowed.

5. *filename* must be enclosed in double quotes if *filename* contains blanks or other special characters.

Menu

Statistics > Multiple imputation

Description

mi append is append for mi data; see [D] **append** for a description of appending datasets.

Options

generate(*newvar*) specifies that new variable *newvar* be created containing 0 for observations from the master and 1 for observations from the using.

nolabel prevents copying the value-label definitions from the using data to the master. Even if you do not specify this option, label definitions from the using never replace those of the master.

nonotes prevents any notes in the using from being incorporated into the master; see [D] **notes**.

force allows string variables to be appended to numeric variables and vice versa. The results of such type mismatches are, of course, missing values. Default behavior is to issue an error message rather than append datasets with such violently differing types.

noupdate in some cases suppresses the automatic mi update this command might perform; see [MI] **noupdate option**.

Remarks

Use mi append when you would use append if the data were not mi.

Remarks are presented under the following headings:

> Adding new observations
> Adding new observations and imputations
> Adding new observations and imputations, M unequal
> Treatment of registered variables

Adding new observations

Assume that file mymi.dta contains data on three-quarters of the patients in the ICU. The data are mi set and $M = 5$. File remaining.dta arrives containing the remaining patients. The data are not mi set. To combine the datasets, you type

```
. use mymi, clear
. append using remaining
```

The original mi data had $M = 5$ imputations, and so do the resulting data. If the new data contain no missing values of the imputed variables, you are ready to go. Otherwise, you will need to impute values for the new data.

Adding new observations and imputations

Assume that file westwing.dta contains data on patients in the west wing of the ICU. File eastwing.dta contains data on patients in the east wing of the ICU. Both datasets are mi set with $M = 5$. You originally intended to analyze the datasets separately, but you now wish to combine them. You type

```
. use westwing, clear
. append using eastwing
```

The original data had $M = 5$ imputations, and so do the resulting data.

The data for $m = 0$ are the result of running an ordinary append on the two $m = 0$ datasets.

The data for $m = 1$ are also the result of running an ordinary append, this time on the two $m = 1$ datasets. Thus the result is a combination of observations of westwing.dta and eastwing.dta in the same way that $m = 0$ is. Imputations for observations that previously existed are obtained from westwing.dta, and imputations for the newly appended observations are obtained from eastwing.dta.

Adding new observations and imputations, M unequal

Consider the same situation as above, but this time assume $M = 5$ in westwing.dta and $M = 4$ in eastwing.dta. The combined result will still have $M = 5$. Imputed values in $m = 5$ will be missing for imputed variables.

Treatment of registered variables

It is possible that the two datasets will have variables registered inconsistently.

Variables registered as imputed in either dataset will be registered as imputed in the final result regardless of how they were registered (or unregistered) in the other dataset.

Barring that, variables registered as passive in either dataset will be registered as passive in the final result.

Barring that, variables registered as regular in either dataset will be registered as regular in the final result.

Saved results

mi append saves the following in r():

Scalars
    r(N_master) number of observations in $m=0$ in master
    r(N_using)   number of observations in $m=0$ in using
    r(M_master) number of imputations (M) in master
    r(M_using)   number of imputations (M) in using

Macros
    r(newvars)   new variables added

Thus values in the resulting data are

N = # of observations in $m = 0$
   = r(N_master) + r(N_using)

k = # of variables
   = k_master + `:word count `r(newvars)''

M = # of imputations
    = max(r(M_master), r(M_using))

Also see

[MI] **intro** — Introduction to mi

[D] **append** — Append datasets

[MI] **mi add** — Add imputations from another mi dataset

[MI] **mi merge** — Merge mi data

Title

mi convert — Change style of mi data

Syntax

mi convert <u>w</u>ide	[, *options*]
mi convert <u>ml</u>ong	[, *options*]
mi convert <u>fl</u>ong	[, *options*]
mi convert <u>fl</u>ongsep *name*	[, *options*]

options	description
clear	okay to convert if data not saved
<u>nou</u>pdate	see [MI] **noupdate option**

Menu

Statistics > Multiple imputation

Description

mi convert converts mi data from one style to another.

Options

clear specifies that it is okay to convert the data even if the data have not been saved to disk since they were last changed.

noupdate in some cases suppresses the automatic mi update this command might perform; see [MI] **noupdate option**.

Remarks

Remarks are presented under the following headings:

> Using mi convert as a convenience tool
> Converting from flongsep
> Converting to flongsep

Using mi convert as a convenience tool

Some tasks are easier in one style than another. `mi convert` allows you to switch to the more convenient style. It would not be unreasonable for a snippet of a session to read

```
. mi convert wide
. drop if sex=="male"
. mi convert mlong, clear
. replace age2 = age^2
```

This user is obviously exploiting his or her knowledge of [MI] **styles**. The official way to do the above would be

```
. drop if sex=="male"
. mi update
. mi passive: replace age2 = age^2
```

It does not matter which approach you choose.

Converting from flongsep

If you have flongsep data, it is worth finding out whether you can convert it to one of the other styles. The other styles are more convenient than flongsep, and `mi` commands run faster on them. With your flongsep data in memory, type

```
. mi convert mlong
```

The result will be either success or an insufficient-memory error. If you get an error, you will have to either live with the flongsep style or increase the amount of memory allocated to Stata; see [D] **memory**. To increase the amount of memory, type

```
. clear all
. set memory 50m          ← (for instance)
. use your_flongsep_data
. mi convert mlong
```

Our use of 50m is just an example. You need to choose a number large enough to hold the data but not so large as to cause your computer to use virtual memory, because then everything runs slowly.

You can make a crude guess as to how much memory is required, as follows:

1. Use your flongsep data. Type `mi describe`. Write down M, the number of imputations, and write down the number of complete observations, which we will call N, and the number of incomplete observations, which we will call n.

2. With your flongsep data still in memory, type `memory`. Write down the number reported as "data + overhead", which we will call S for size.

3. Calculate $T = S + M \times S \times (n/N)$. T is an approximation of the memory your `mi` data would consume in the mlong style. To that, we need to add a bit to account for extra memory used by Stata commands, and for variables or observations you might want to add. How much to add is always debatable. For large datasets, add 10% or 5 MB, whichever is smaller.

For instance, you might have

$$\begin{aligned} M &= 30 \\ N &= 10{,}000 \\ n &= 1{,}500 \\ S &= 8{,}040{,}000 = 8 \text{ MB} \end{aligned}$$

and thus we would calculate $T = 8 + 30 \times 8 \times (1500/10000) = 44$ MB, to which we would add another 4 or 5 MB, to obtain 48 or 49 MB. So we would try `set memory 49m`.

Converting to flongsep

Note that `mi convert`'s syntax for converting to flongsep is

`mi convert flongsep` *name*

You must specify a name, and that name will become the basis for the names of the datasets that comprise the collection of flongsep data. Data for $m = 0$ will be stored in *name*.dta; data for $m = 1$, in _1_*name*.dta; data for $m = 2$, in _2_*name*.dta; and so on. The files will be stored in the current directory; see the `pwd` command in [D] **cd**.

If you are going to use flongsep data, see *Advice for using flongsep* in [MI] **styles**. Also see [MI] **mi copy** and [MI] **mi erase**.

Also see

[MI] **intro** — Introduction to mi

[MI] **styles** — Dataset styles

Title

> **mi copy** — Copy mi flongsep data

Syntax

> mi copy *newname* [, replace]

Menu

Statistics > Multiple imputation

Description

mi copy *newname* copies flongsep data in memory to *newname* and sets it so that you are working with that copy. *newname* may not be specified with the .dta suffix.

In detail, mi copy *newname* 1) completes saving the flongsep data to its current name if that is necessary; 2) copies the data to *newname*.dta, _1_*newname*.dta, _2_*newname*.dta, ..., _M_*newname*.dta; and 3) tells mi that you are now working with *newname*.dta in memory.

mi copy can also be used with wide, mlong, or flong data, although there is no reason you would want to do so. The data are not saved to the original filename as flongsep data would be, but otherwise actions are the same: the data in memory are copied to *newname*.dta, and *newname*.dta is loaded into memory.

Option

replace specifies that it is okay to overwrite *newname*.dta, _1_*newname*.dta, _2_*newname*.dta, ..., if they already exist.

Remarks

In Stata, one usually works with a copy of the data in memory. Changes you make to the data are not saved in the underlying disk file until and unless you explicitly save your data. That is not true when working with flongsep data.

Flongsep data are a matched set of datasets, one containing $m = 0$, another containing $m = 1$, and so on. You work with one of them in memory, namely, $m = 0$, but as you work, the other datasets are automatically updated; as you make changes, the datasets on disk change.

Therefore, it is best to work with a copy of your flongsep data and then periodically save the data to the real files, thus mimicking how you work with ordinary Stata datasets. mi copy is for just that purpose. After loading your flongsep data, type, for example,

    . use myflongsep

and immediately make a copy,

    . mi copy *newname*

You are now working with the same data but under a new name. Your original data are safe.

When you reach a point where you would ordinarily save your data, whether under the original name or a different one, type

. mi copy *originalname_or_different_name*, replace

. use *newname*

Later, when you are done with *newname*, you can erase it by typing

. mi erase *newname*

Concerning erasure, you will discover that `mi erase` will not let you erase the files when you have one of the files to be erased in memory. Then you will have to type

. mi erase *newname*, clear

See [MI] **mi erase** for more information.

For more information on flongsep data, see *Advice for using flongsep* in [MI] **styles**.

Also see

[MI] **intro** — Introduction to mi

[MI] **styles** — Dataset styles

[MI] **mi erase** — Erase mi datasets

Title

> **mi describe** — Describe mi data

Syntax

    mi <u>q</u>uery

    mi <u>d</u>escribe [, *describe_options*]

describe_options	description
<u>det</u>ail	show missing-value counts for $m = 1$, $m = 2$, ...
<u>noup</u>date	see [MI] **noupdate option**

Menu

Statistics > Multiple imputation

Description

mi query reports whether the data in memory are mi data and, if they are, reports the style in which they are set.

mi describe provides a more detailed report on mi data.

Options

detail reports the number of missing values in $m = 1$, $m = 2$, ..., $m = M$ in the imputed and passive variables, along with the number of missing values in $m = 0$.

noupdate in some cases suppresses the automatic mi update this command might perform; see [MI] **noupdate option**.

Remarks

Remarks are presented under the following headings:

    mi query
    mi describe

mi query

mi query without mi data in memory reports

```
. mi query
(data not mi set)
```

With mi data in memory, you see something like

```
. mi query
data mi set wide, M = 15
last mi update 16apr2009 09:57:53, approximately 5 minutes ago
```

mi query does not burden you with unnecessary information. It mentions when mi update was last run because you should run it periodically; see [MI] **mi update**.

mi describe

mi describe more fully describes mi data:

```
. mi describe
Style:  mlong
        last mi update 16apr2009 10:21:07, approximately 2 minutes ago
Obs.:   complete         90
        incomplete       10    (M = 20 imputations)
        ----------------------
        total           100
Vars.:  imputed: 2; smokes(10) age(5)
        passive: 1; agesq(5)
        regular: 0
        system:  3; _mi_m _mi_id _mi_miss
        (there are 3 unregistered variables; gender race chd)
```

mi describe lists the style of the data, the number of complete and incomplete observations, M (the number of imputations), the registered variables, and the number of missing values in $m = 0$ of the imputed and passive variables. In the output, the line

```
Vars.:  imputed: 2; smokes(10) age(5)
```

means that the smokes variable contains 10 missing values in $m = 0$ and that age contains 5. Those values are soft missings and thus eligible to be imputed. If one of smokes' missing values in $m = 0$ were hard, the line would read

```
Vars.:  imputed: 2; smokes(9+1) age(5)
```

mi describe reports information about $m = 0$. To obtain information about all m's, use mi describe, detail:

```
. mi describe, detail
Style:  mlong
        last mi update 16apr2009 10:36:50, approximately 3 minutes ago
Obs.:   complete         90
        incomplete       10    (M = 20 imputations)
        ----------------------
        total           100
Vars.:  imputed: 2; smokes(10; 20*0) age(5; 20*0)
        passive: 1; agesq(5; 20*0)
        regular: 0
        system:  3; _mi_m _mi_id _mi_miss
        (there are 3 unregistered variables; gender race chd)
```

In this example, all imputed values are nonmissing. We can see that from

```
Vars.:  imputed: 2; smokes(10; 20*0) age(5; 20*0)
```

Note the 20*0 after the semicolons. That is the number of missing values in $m=1$, $m=2$, ..., $m=20$. In the smokes variable, there are 10 missing values in $m=0$, then 0 in $m=1$, then 0 in $m=2$, and so on. If $m=17$ had two missing imputed values, the line would read

    Vars.: imputed: 2; smokes(10; 16*0, 2, 3*0) age(5; 20*0)

16*0, 2, 3*0 means that for $m=1$, $m=2$, ..., $m=20$, the first 16 have 0 missing values, the next has 2, and the last 3 have 0.

If smokes had $9+1$ missing values rather than 10, i.e, 9 soft missing values plus 1 hard missing rather than all 10 being soft missing, and all 9 soft missings were filled in, the line would read

    Vars.: imputed: 2; smokes(9+1; 20*0) age(5; 20*0)

The 20 imputations are shown as having no soft missing values. It goes without saying that they have 1 hard missing. Think of 20*0 as meaning 20*(0+1).

If smokes had $9+1$ missing values and two of the soft missings in $m=18$ were still missing, the line would read

    Vars.: imputed: 2; smokes(9+1; 16*0, 2, 3*0) age(5; 20*0)

Saved results

mi query saves the following in r():

   Scalars
       r(update)   seconds since last mi update
       r(m) m if r(style)=="flongsep"
       r(M) M if r(style)!="flongsep"
   Macros
       r(style)   style
       r(name)   name if r(style)=="flongsep"

Note that mi query issues a return code of 0 even if the data are not mi. In that case, r(style) is "".

mi describe saves the following in r():

   Scalars
       r(update)   seconds since last mi update
       r(N)   number of observations in $m=0$
       r(N_incomplete)   number of incomplete observations in $m=0$
       r(N_complete)   number of complete observations in $m=0$
       r(M) M
   Macros
       r(style)   style
       r(ivars)   names of imputed variables
       r(_0_miss_ivars)   #=. in each r(ivar) in $m=0$
       r(_0_hard_ivars)   #>. in each r(ivar) in $m=0$
       r(pvars)   names of passive variables
       r(_0_miss_pvars)   #≥. in each r(pvar) in $m=0$
       r(rvars)   names of regular variables

If the detail option is specified, for each m, $m = 1, 2, \ldots, M$, also saved are

> Macros
> r(_m_miss_ivars)   #=. in each r(ivar) in m
> r(_m_miss_pvars)   #≥. in each r(pvar) in m

Also see

[MI] **intro** — Introduction to mi

Title

mi erase — Erase mi datasets

Syntax

mi erase *name* [, clear]

Menu

Statistics > Multiple imputation

Description

mi erase erases mi .dta datasets.

Option

clear specifies that it is okay to erase the files even if one of the files is currently in memory. If clear is specified, the data are dropped from memory and the files are erased.

Remarks

Stata's ordinary erase (see [D] **erase**) is not sufficient for erasing mi datasets because an mi dataset might be flongsep, in which case the single name would refer to a collection of files, one containing $m = 0$, another containing $m = 1$, and so on. mi erase deletes all the files associated with mi dataset *name*.dta, which is to say, it erases *name*.dta, _1_*name*.dta, _2_*name*.dta, and so on:

```
. mi erase mysep
(files mysep.dta, _1_mysep.dta _2_mysep.dta _3_mysep.dta erased)
```

Also see

[MI] **intro** — Introduction to mi

[MI] **styles** — Dataset styles

[MI] **mi copy** — Copy mi flongsep data

Title

mi estimate — Estimation using multiple imputations

Syntax

Compute MI estimates of coefficients by fitting estimation command to mi data

    mi <u>est</u>imate [, *options*] : *estimation_command* ...

Compute MI estimates of transformed coefficients by fitting estimation command to mi data

    mi <u>est</u>imate [*spec*] [, *options*] : *estimation_command* ...

where *spec* may be one or more terms of the form ([*name*:] *exp*). *exp* is any function of the parameter estimates allowed by nlcom; see [R] **nlcom**.

options	description
Options	
<u>ni</u>mputations(*#*)	specify number of imputations to use; default is to use all existing imputations
<u>im</u>putations(*numlist*)	specify which imputations to use
<u>ufmit</u>est	perform unrestricted FMI model test
nosmall	do not apply small-sample adjustment to degrees of freedom
<u>sav</u>ing(*miestfile*[, replace])	save individual estimation results to *miestfile*.ster
Tables	
[<u>no</u>]<u>cit</u>able	suppress/display standard estimation table containing parameter-specific confidence intervals; default is citable
<u>dft</u>able	display degrees-of-freedom table; dftable implies nocitable
<u>var</u>table	display variance information about estimates; vartable implies citable
table_options	control table output
display_options	control spacing and display of omitted variables and base and empty cells
Reporting	
<u>level</u>(*#*)	set confidence level; default is level(95)
dots	display dots as estimations performed
<u>noi</u>sily	display any output from *estimation_command* (and from nlcom if transformations specified)
trace	trace *estimation_command* (and nlcom if transformations specified); implies noisily

Advanced

esample(*newvar*)	store estimation sample in variable *newvar*; available only in the flong and flongsep styles
errorok	allow estimation even when *estimation_command* (or nlcom) errors out; such imputations are discarded from the analysis
esampvaryok	allow estimation when estimation sample varies across imputations
cmdok	allow estimation when *estimation_command* is not one of the officially supported estimation commands
†<u>coef</u>legend	display coefficients' legend instead of coefficient table; implies nocitable and cannot be specified with citable or dftable
†<u>nowarn</u>ing	suppress the warning about varying estimation samples
†*eform_option*	display coefficient table in exponentiated form
†post	post estimated coefficients and VCE to e(b) and e(V)
†<u>noupdate</u>	do not perform mi update; see [MI] **noupdate option**

† coeflegend, nowarning, *eform_option*, post, and noupdate do not appear in the dialog box.

You must mi set your data before using mi estimate; see [MI] **mi set**.

table_options	description
<u>nohe</u>ader	suppress table header(s)
<u>nota</u>ble	suppress table(s)
nocoef	suppress table output related to coefficients
<u>nocmd</u>legend	suppress command legend that appears in the presence of transformed coefficients when nocoef is used
<u>notr</u>coef	suppress table output related to transformed coefficients
<u>nole</u>gend	suppress table legend(s)
<u>nocns</u>report	do not display constraints

See [MI] **mi estimate postestimation** for features available after estimation. mi estimate is its own estimation command. The postestimation features for mi estimate do not include by default the postestimation features for *estimation_command*. To replay results, type mi estimate without arguments.

Menu

Statistics > Multiple imputation

Description

mi estimate: *estimation_command* runs *estimation_command* on the imputed mi data, and adjusts coefficients and standard errors for the variability between imputations according to the combination rules by Rubin (1987).

Options

Options

nimputations(#) specifies that the first # imputations be used; # must be $2 \leq \# \leq M$. The default is to use all imputations, M. Only one of nimputations() or imputations() may be specified.

imputations(*numlist*) specifies which imputations to use. The default is to use all of them. *numlist* must contain at least two numbers. Only one of nimputations() or imputations() may be specified.

ufmitest specifies that the unrestricted fraction missing information (FMI) model test be used. The default test performed assumes equal fractions of information missing due to nonresponse for all coefficients. This is equivalent to the assumption that the between-imputation and within-imputation variances are proportional. The unrestricted test may be preferable when this assumption is suspect provided the number of imputations is large relative to the number of estimated coefficients.

nosmall specifies that no small-sample adjustment be made to the degrees of freedom. By default, individual tests of coefficients (and transformed coefficients) use the small-sample adjustment of Barnard and Rubin (1999) and the overall model test uses the small-sample adjustment of Reiter (2007).

saving(*miestfile*[, replace]) saves estimation results from each model fit in *miestfile*.ster. The replace suboption specifies to overwrite *miestfile*.ster if it exists. *miestfile*.ster can later be used by mi estimate using (see [MI] **mi estimate using**) to obtain MI estimates of coefficients or of transformed coefficients without refitting the completed-data models. This file is written in the format used by estimates use; see [R] **estimates save**.

Tables

All table options below may be specified at estimation time or when redisplaying previously estimated results. Table options must be specified as options to mi estimate, not to *estimation_command*.

citable and nocitable specify whether the standard estimation table containing parameter-specific confidence intervals is displayed. The default is citable. nocitable can be used with vartable to suppress the confidence-interval table.

dftable displays a table containing parameter-specific degrees of freedom and percentages of increase in standard errors due to nonresponse. dftable implies nocitable.

vartable displays a table reporting variance information about MI estimates. The table contains estimates of within-imputation variances, between-imputation variances, total variances, relative increases in variance due to nonresponse, fractions of information about parameter estimates missing due to nonresponse, and relative efficiencies for using finite M rather than a hypothetically infinite number of imputations. vartable implies citable.

table_options control the appearance of all displayed table output:

noheader suppresses all header information from the output. The table output is still displayed.

notable suppresses all tables from the output. The header information is still displayed.

nocoef suppresses the display of tables containing coefficient estimates. This option affects the table output produced by citable, dftable, and vartable.

nocmdlegend suppresses the table legend showing the specified command line, *estimation_command*, from the output. This legend appears above the tables containing transformed coefficients (or above the variance-information table if vartable is used) when nocoef is specified.

notrcoef suppresses the display of tables containing estimates of transformed coefficients (if specified). This option affects the table output produced by citable, dftable, and vartable.

nolegend suppresses all table legends from the output.

nocnsreport; see [R] **estimation options**.

display_options: noomitted, vsquish, noemptycells, baselevels, allbaselevels; see [R] **estimation options**.

⌐ Reporting ⌐

Reporting options must be specified as options to mi estimate and not as options to *estimation_command*.

level(*#*); see [R] **estimation options**.

dots specifies that dots be displayed as estimations are successfully completed. An **x** is displayed if the *estimation_command* returns an error, if the model fails to converge, or if nlcom fails to estimate one of the transformed coefficients specified in *spec*.

noisily specifies that any output from *estimation_command* and nlcom, used to obtain the estimates of transformed coefficients if transformations are specified, be displayed.

trace traces the execution of *estimation_command* and traces nlcom if transformations are specified. trace implies noisily.

⌐ Advanced ⌐

esample(*newvar*) creates *newvar* containing e(sample). This option is useful to identify which observations were used in the estimation, especially when the estimation sample varies across imputations (see *Potential problems that can arise when using mi estimate* for details). *newvar* is zero in the original data ($m = 0$) and in any imputations ($m > 0$) in which the estimation failed or that were not used in the computation. esample() may be specified only if the data are flong or flongsep; see [MI] **mi convert** to convert to one of those styles. The variable created will be super varying and therefore must not be registered; see [MI] **mi varying** for more explanation. The saved estimation sample *newvar* may be used later with mi extract (see [MI] **mi extract**) to set the estimation sample.

errorok specifies that estimations that fail be skipped and the combined results be based on the successful individual estimation results. The default is that mi estimate stops if an individual estimation fails. If errorok is specified with saving(), all estimation results, including failed, are saved to a file.

esampvaryok allows estimation to continue even if the estimation sample varies across imputations. mi estimate stops if the estimation sample varies. If esampvaryok is specified, results from all imputations are used to compute MI estimates and a warning message is displayed at the bottom of the table. Also see the esample() option. See *Potential problems that can arise when using mi estimate* for more information.

cmdok allows unsupported estimation commands to be used with mi estimate; see [MI] **estimation** for a list of supported estimation commands. Alternatively, if you want mi estimate to work with your estimation command, add the property mi to the program properties; see [P] **program properties**.

The following options are available with mi estimate but are not shown in the dialog box:

coeflegend; see [R] **estimation options**. coeflegend implies nocitable and cannot be combined with citable or dftable.

nowarning suppresses the warning message at the bottom of table output that occurs if the estimation sample varies and esampvaryok is specified. See *Potential problems that can arise when using mi estimate* for details.

eform_option; see [R] **eform_option**. Regardless of the *estimation_command* specified, mi estimate reports results in the coefficient metric under which the combination rules are applied. You may use the appropriate *eform_option* to redisplay results in exponentiated form, if desired. If dftable is also specified, the reported degrees of freedom and percentage increases in standard errors are not adjusted and correspond to the original coefficient metric.

post requests MI estimates of coefficients and their respective VCEs be posted in the usual way. This allows the use of *estimation_command*-specific postestimation tools with MI estimates. There are issues; see *Using the command-specific postestimation tools* in [MI] **mi estimate postestimation**. post may be specified at estimation time or when redisplaying previously estimated results.

noupdate in some cases suppresses the automatic mi update this command might perform; see [MI] **noupdate option**. This option is seldom used.

Remarks

mi estimate requires that imputations be already formed; see [MI] **mi impute**. To import existing multiply imputed data, see [MI] **mi import**.

Remarks are presented under the following headings:

>Using mi estimate
>Example 1: Completed-data logistic analysis
>Example 2: Completed-data regression analysis
>Example 3: Completed-data survival analysis
>Example 4: Estimating transformations
>Potential problems that can arise when using mi estimate

Using mi estimate

mi estimate estimates model parameters from multiply imputed data and adjusts coefficients and standard errors for the variability between imputations. It runs the specified *estimation_command* on each of the M imputed datasets to obtain the M completed-data estimates of coefficients and their VCEs. It then computes MI estimates of coefficients and standard errors by applying combination rules (Rubin 1987, 77) to the M completed-data estimates. See [MI] **intro substantive** for a discussion of MI analysis and see *Methods and formulas* for computational details.

To use mi estimate, your data must contain at least two imputations. The basic syntax of mi estimate is

. mi estimate: *estimation_command* ...

estimation_command is any estimation command from the list of supported estimation commands; see [MI] **estimation**.

If you wish to estimate on survey data, type

. mi estimate: svy: *estimation_command* ...

If you want to vary the number of imputations or select which imputations to use in the computations, use the nimputations() or the imputations() option, respectively.

. mi estimate, nimputations(9): *estimation_command* ...

Doing so is useful to evaluate the stability of MI results.

You can obtain more-detailed information about imputation results by specifying the dftable and vartable options.

You can additionally obtain estimates of transformed coefficients by specifying expressions with mi estimate; see *Example 4: Estimating transformations* for details.

When using mi estimate, keep in mind that

1. mi estimate is its own estimation command.

2. mi estimate uses different degrees of freedom for each estimated parameter when computing its significance level and confidence interval.

3. mi estimate reports results in the coefficient metric under which combination rules are applied regardless of the default reporting metric of the specified *estimation_command*. Use *eform_option* with mi estimate to report results in the exponentiated metric, if you wish.

4. mi estimate has its own reporting options and does not respect command-specific reporting options. The reporting options specified with *estimation_command* affect only the output of the command that is displayed when mi estimate's noisily option is specified. Specify mi estimate's options immediately after the mi estimate command:

    . mi estimate, *options*: *estimation_command* ...

Example 1: Completed-data logistic analysis

Recall the logistic analysis of the heart attack data from [MI] **intro substantive**. The goal of the analysis was to explore the relationship between heart attacks and smoking adjusted for other factors such as age, body mass index, gender, and educational status. The original data contain missing values of BMI. The listwise-deletion analysis on the original data determined that smoking and BMI have significant impact on a heart attack. After imputing missing values of BMI, age was determined to be a significant factor as well. See *A brief introduction to MI using Stata* in [MI] **intro substantive** for details. The data we used are stored in mheart1s20.dta.

Below we refit the logistic model using the imputed data. We also specify the dots option so that dots will be displayed as estimations are completed.

(*Continued on next page*)

```
. use http://www.stata-press.com/data/r11/mheart1s20
(Fictional heart attack data; bmi missing)
. mi estimate, dots: logit attack smokes age bmi hsgrad female
Imputations (20):
.........10.........20 done
```

Multiple-imputation estimates				Imputations		=	20
Logistic regression				Number of obs		=	154
				Average RVI		=	0.0312
DF adjustment:	Large sample			DF:	min	=	1060.38
					avg	=	223362.56
					max	=	493335.88
Model F test:	Equal FMI			F(5,71379.3)		=	3.59
Within VCE type:	OIM			Prob > F		=	0.0030

attack	Coef.	Std. Err.	t	P>|t|	[95% Conf. Interval]	
smokes	1.198595	.3578195	3.35	0.001	.4972789	1.899911
age	.0360159	.0154399	2.33	0.020	.0057541	.0662776
bmi	.1039416	.0476136	2.18	0.029	.010514	.1973692
hsgrad	.1578992	.4049257	0.39	0.697	-.6357464	.9515449
female	-.1067433	.4164735	-0.26	0.798	-.9230191	.7095326
_cons	-5.478143	1.685075	-3.25	0.001	-8.782394	-2.173892

The left header column reports information about the fitted MI model. The right header column reports the number of imputations and the number of observations used, the average relative variance increase (RVI) due to nonresponse, a summary about parameter-specific degrees of freedom (DF), and the overall model test that all coefficients, excluding the constant, are equal to zero.

Notice first that mi estimate reports Student's t and F statistics for inference although logit would usually report Z and χ^2 statistics.

mi estimate: logit is not logit. mi estimate uses Rubin's combination rules to obtain the estimates from multiply imputed data. The variability of the MI estimates consists of two components: variability within imputations and variability between imputations. Therefore, the precision of the MI estimates is governed not only by the number of observations in the sample but also by the number of imputations. As such, even if the number of observations is large, if the number of imputations is small, the reference distribution used for inference will deviate from the normal distribution. Because in practice the number of imputations tends to be small, mi estimate uses a reference t distribution.

Returning to the output, average RVI reports the average relative increase (averaged over all coefficients) in variance of the estimates because of the missing bmi values. A relative variance increase is an increase in the variance of the estimate because of the loss of information about the parameter due to nonresponse relative to the variance of the estimate with no information lost. The closer this number is to zero, the less effect missing data have on the variance of the estimate. Note that the reported RVI will be zero if you use mi estimate with the complete data or with missing data that have not been imputed. In our case, average RVI is small: 0.03.

The coefficient-specific degrees of freedom (DF) averaging 223,363 are large. They are large because the MI degrees of freedom depends not only on the number of imputations but also on the RVI due to nonresponse. Specifically, the degrees of freedom is inversely related to RVI. The closer RVI is to zero, the larger the degrees of freedom regardless of the number of imputations.

To the left of the DF, we see that the degrees of freedom are obtained under a large-sample assumption. The alternative is to use a small-sample adjustment. Whether the small-sample adjustment is applied is determined by the type of the reference distribution used for inference by the specified estimation command. For the commands that use a large-sample (normal) approximation for inference, a large-sample approximation is used when computing the MI degrees of freedom. For the commands that use a small-sample (Student's t) approximation for inference, a small-sample approximation is used when computing the MI degrees of freedom. See *Methods and formulas* for details. As we already mentioned, `logit` assumes large samples for inference, and thus the MI degrees of freedom is computed assuming a large sample.

The model F test rejects the hypothesis that all coefficients are equal to zero and thus rules out a constant-only model for heart attacks. By default, the model test uses the assumption that the fractions of missing information of all coefficients are equal (as noted by `Equal FMI` to the left). Although this assumption may not be supported by the data, it is used to circumvent the difficulties arising with the estimation of the between-imputation variance matrix based on a small number of imputations. See *Methods and formulas* and [MI] **mi estimate postestimation** for details.

`mi estimate` also reports the type of variance estimation used by the estimation command to compute variance estimates in the individual completed-data analysis. These completed-data variance estimates are then used to compute the within-imputation variance. In our example, the observed-information-matrix (OIM) method, the default variance-estimation method used by maximum likelihood estimation, is used to compute completed-data VCEs. This is labeled as `Within VCE type: OIM` in the output.

Finally, `mi estimate` reports a coefficient table containing the combined estimates. Unlike all other Stata estimation commands, the reported significance levels and confidence intervals in this table are based on degrees of freedom that is specific to each coefficient. Remember that the degrees of freedom depends on the relative variance increases and thus on how much information is lost about the estimated parameter due to missing data. How much information is lost is specific to each parameter and so is the degrees of freedom.

As we already saw, a summary of the coefficient-specific degrees of freedom (minimum, average, and maximum) was reported in the header. We can obtain a table containing coefficient-specific degrees of freedom by replaying the results with the `dftable` option:

```
. mi estimate, dftable

Multiple-imputation estimates              Imputations      =         20
Logistic regression                        Number of obs    =        154
                                           Average RVI      =     0.0312
DF adjustment:    Large sample             DF:      min     =    1060.38
                                                    avg     =  223362.56
                                                    max     =  493335.88
Model F test:       Equal FMI              F(  5,71379.3)   =       3.59
Within VCE type:         OIM               Prob > F         =     0.0030
```

						% Increase
attack	Coef.	Std. Err.| t	P>|t|| DF	Std. Err.
smokes	1.198595	.3578195	3.35	0.001	320019.4	0.39
age	.0360159	.0154399	2.33	0.020	493335.9	0.31
bmi	.1039416	.0476136	2.18	0.029	1060.4	7.45
hsgrad	.1578992	.4049257	0.39	0.697	165126.7	0.54
female	-.1067433	.4164735	-0.26	0.798	358078.3	0.37
_cons	-5.478143	1.685075	-3.25	0.001	2554.8	4.61

Notice that we type `mi estimate` to replay the results, not `logit`.

The header information remains the same. In particular, degrees of freedom ranges from 1,060 to 493,336 and averages 223,363. In the table output, the columns for the confidence intervals are replaced with the `DF` and `% Increase Std. Err.` columns. We now see that the smallest degrees of freedom corresponds to the coefficient for `bmi`. We should have anticipated this because `bmi` is the only variable containing missing values in this example. The largest degrees of freedom is observed for the coefficient for `age`, which suggests that the loss of information due to nonresponse is the smallest for the estimation of this coefficient.

The last column displays as percentages the increase in standard errors of the parameters due to nonresponse. We observe a 7% increase in the standard error for the coefficient of `bmi` and a 4% increase in the standard error for the constant. Increases in standard errors of other coefficients are negligible.

In this example, we displayed a degrees-of-freedom table on replay by specifying the `dftable` option. We could also obtain this table if we specified this option at estimation time. Alternatively, if desired, we could display both tables by specifying the `citable` and `dftable` options together.

We can obtain more detail about imputation results by specifying the `vartable` option. We specify this option on replay and also use the `nocitable` option to suppress the default confidence-interval table:

```
. mi estimate, vartable nocitable
Multiple-imputation estimates                   Imputations     =          20
Variance information
```

	Imputation variance			RVI	FMI	Relative efficiency
	Within	Between	Total			
smokes	.127048	.00094	.128035	.007765	.007711	.999615
age	.000237	1.4e-06	.000238	.006245	.00621	.99969
bmi	.001964	.000289	.002267	.154545	.135487	.993271
hsgrad	.162206	.001675	.163965	.010843	.010739	.999463
female	.172187	.001203	.17345	.007338	.00729	.999636
_cons	2.5946	.233211	2.83948	.094377	.086953	.995671

The first three columns of the table provide the variance information specific to each parameter. As we already discussed, MI variance contains two sources of variation: within imputation and between imputation. The first two columns provide estimates for the within-imputation and between-imputation variances. The third column is a total variance that is the sum of the two variances plus an adjustment for using a finite number of imputations. The next two columns are individual RVIs and fractions of missing information (FMIs) due to nonresponse. The last column records relative efficiencies for using a finite number of imputations (20 in our example) versus the theoretically optimal infinite number of imputations.

We notice that the coefficient for `age` has the smallest within-imputation and between-imputation variances. The between-imputation variability is very small relative to the within-imputation variability, which is why `age` had such a large estimate of the degrees of freedom we observed earlier. Correspondingly, this coefficient has the smallest values for RVI and FMI. As expected, the coefficient for `bmi` has the highest RVI and FMI.

The reported relative efficiencies are high for all coefficient estimates, with the smallest relative efficiency, again, corresponding to `bmi`. These estimates, however, are only approximations and thus should not be used exclusively to determine the required number of imputations. See Royston, Carlin, and White (2009) for other ways of determining a suitable number of imputations.

Example 2: Completed-data regression analysis

Recall the data on house resale prices from example 3 of [MI] **mi impute mvn**. We use the imputed data stored in `mhouses1993s30.dta` to examine the relationship of various predictors on price via linear regression:

```
. use http://www.stata-press.com/data/r11/mhouses1993s30
(Albuquerque Home Prices Feb15-Apr30, 1993)

. mi estimate, ni(5): regress price tax sqft age nfeatures ne custom corner
```

Multiple-imputation estimates   Imputations   =   5
Linear regression   Number of obs   =   117
                                    Average RVI   =   0.2939
                                    Complete DF   =   109
DF adjustment: Small sample   DF: min   =   7.51
                                             avg   =   61.46
                                             max   =   94.51
Model F test: Equal FMI   F(7,   79.3) =   53.00
Within VCE type: OLS   Prob > F   =   0.0000

price	Coef.	Std. Err.	t	P>|t|	[95% Conf. Interval]	
tax	.5375429	.1399677	3.84	0.001	.2491601	.8259257
sqft	.2940729	.0890706	3.30	0.004	.1066042	.4815415
age	-.7325252	.7605651	-0.96	0.365	-2.506304	1.041254
nfeatures	3.696554	13.69565	0.27	0.788	-23.62366	31.01677
ne	8.303979	34.64444	0.24	0.811	-60.47857	77.08653
custom	129.2014	44.29287	2.92	0.004	41.25603	217.1468
corner	-63.77848	41.48302	-1.54	0.128	-146.1749	18.61793
_cons	135.4265	64.57713	2.10	0.039	7.184164	263.6688

By default, mi estimate uses all available imputations in the analysis. For the purpose of illustration, we use only the first 5 imputations out of the available 30 by specifying the nimputations(5) option, which we abbreviated as ni(5).

Compared with the output from the previous example, an additional result, Complete DF, is reported. Also notice that the adjustment for the degrees of freedom is now labeled as Small sample. Remember that mi estimate determines what adjustment to use based on the reference distribution used for inference by the specified estimation command.

regress uses a reference t distribution with $117 - 8 = 109$ residual degrees of freedom. Thus a small-sample adjustment is used by mi estimate for the MI degrees of freedom.

Complete DF contains the degrees of freedom used for inference in each completed dataset (imputation). It corresponds to the completed-data residual degrees of freedom saved by the command in e(df_r). In most applications, the completed-data residual degrees of freedom will be the same, and so Complete DF will correspond to the *complete degrees of freedom*, the degrees of freedom that would have been used for inference if the data were complete. In the case when the completed-data residual degrees of freedom varies across imputations (as may happen when the estimation sample varies; see *Potential problems that can arise when using mi estimate*), Complete DF reports the smallest of them.

In our example, all completed-data residual degrees of freedom are equal, and Complete DF is equal to 109, the completed-data residual degrees of freedom obtained from regress. mi estimate uses the complete degrees of freedom to adjust the MI degrees of freedom for a small sample (Barnard and Rubin 1999).

Example 3: Completed-data survival analysis

Consider survival data on 48 participants in a cancer drug trial. The dataset contains information about participants' ages, treatments received (drug or placebo), times to death measured in months, and a censoring indicator. The data are described in more detail in *Cox regression with censored data* of [ST] **stcox**. We consider a version of these data containing missing values for age. The imputed data are saved in mdrugtrs25.dta:

```
. use http://www.stata-press.com/data/r11/mdrugtrs25
(Patient Survival in Drug Trial)
. mi describe
  Style: mlong
   Obs.: complete          40
         incomplete          8  (M = 25 imputations)
                    ─────────
         total              48
   Vars.: imputed: 1; age(8)
          passive: 0
          regular: 3; studytime died drug
          system:  3; _mi_m _mi_id _mi_miss
         (there are no unregistered variables)
```

The dataset contains 25 imputations for 8 missing values of age.

We analyze these data using stcox with mi estimate. These data have not yet been stset, so we use mi stset (see [MI] **mi XXXset**) to set them and then perform the analysis using mi estimate: stcox:

```
. mi stset studytime, failure(died)
     failure event:  died != 0 & died < .
obs. time interval:  (0, studytime]
 exit on or before:  failure

       48  total obs.
        0  exclusions

       48  obs. remaining, representing
       31  failures in single record/single failure data
      744  total analysis time at risk, at risk from t =         0
                                earliest observed entry t =      0
                                     last observed exit t =     39

. mi estimate, dots: stcox drug age

Imputations (25):
.........10.........20..... done

Multiple-imputation estimates             Imputations     =         25
Cox regression: Breslow method for ties   Number of obs   =         48
                                          Average RVI     =     0.1059
DF adjustment:   Large sample             DF:     min     =     998.63
                                                  avg     =   11621.53
                                                  max     =   22244.42
Model F test:        Equal FMI            F(   2, 4448.6) =      13.43
Within VCE type:          OIM             Prob > F        =     0.0000

─────────────────────────────────────────────────────────────────────
         _t │     Coef.   Std. Err.     t    P>|t|   [95% Conf. Interval]
────────────┼────────────────────────────────────────────────────────
       drug │ -2.204572   .4589047   -4.80   0.000  -3.104057  -1.305086
        age │  .1242711    .040261    3.09   0.002    .0452652   .2032771
─────────────────────────────────────────────────────────────────────
```

Notice that mi estimate displays the results in the coefficient metric and not in the hazard-ratio metric. By default, mi estimate reports results in the metric under which the combination rules were applied. To obtain the results as hazard ratios, we can use the hr option with mi estimate:

```
. mi estimate, hr
Multiple-imputation estimates              Imputations     =         25
Cox regression: Breslow method for ties    Number of obs   =         48
                                           Average RVI     =     0.1059
DF adjustment:   Large sample              DF:      min    =     998.63
                                                    avg    =   11621.53
                                                    max    =   22244.42
Model F test:        Equal FMI             F(   2, 4448.6) =      13.43
Within VCE type:          OIM              Prob > F        =     0.0000
```

_t	Haz. Ratio	Std. Err.	t	P>|t|	[95% Conf. Interval]	
drug	.1102977	.0506161	-4.80	0.000	.0448668	.2711491
age	1.132323	.0455885	3.09	0.002	1.046305	1.225412

We obtain results similar to those from the corresponding example in [ST] **stcox**.

We specified the hr option above on replay. We can also specify it at estimation time:

```
. mi estimate, hr: stcox drug age
  (output omitted)
```

Notice that the hr option must be specified with mi estimate to obtain hazard ratios. Specifying it with the command itself,

```
. mi estimate: stcox drug age, hr
  (output omitted)
```

will not affect the output from mi estimate but only that of the command, stcox. You see stcox's output only if you specify mi estimate's noisily option.

Example 4: Estimating transformations

Stata estimation commands usually support lincom and nlcom (see [R] **lincom** and [R] **nlcom**) to obtain estimates of the transformed coefficients after estimation by using the delta method. Because MI estimates based on a small number of imputations may not yield a valid VCE, this approach is not generally viable. Also, transformations applied to the combined coefficients are only asymptotically equivalent to the combined transformed coefficients. With a small number of imputations, these two ways of obtaining transformed coefficients can differ significantly.

Thus mi estimate provides its own way of combining transformed coefficients. You need to use mi estimate's method for both linear and nonlinear combinations of coefficients. We are about to demonstrate how using the ratio of coefficients as an example, but what we are about to do would be equally necessary if we wanted to obtain the difference in two coefficients.

For the purpose of illustration, suppose that we want to estimate the ratio of the coefficients, say, age and sqft from example 2. We can do this by typing

(Continued on next page)

```
. use http://www.stata-press.com/data/r11/mhouses1993s30, clear
(Albuquerque Home Prices Feb15-Apr30, 1993)

. mi estimate (ratio: _b[age]/_b[sqft]):
> regress price tax sqft age nfeatures ne custom corner

Multiple-imputation estimates              Imputations      =         30
Linear regression                          Number of obs    =        117
                                           Average RVI      =     0.5415
                                           Complete DF      =        109
DF adjustment:    Small sample             DF:      min     =      16.42
                                                    avg     =      72.83
                                                    max     =     101.18
Model F test:         Equal FMI            F(   7,   96.3)  =      45.63
Within VCE type:            OLS            Prob > F         =     0.0000
```

price	Coef.	Std. Err.	t	P>|t|	[95% Conf. Interval]	
tax	.5516444	.1443319	3.82	0.000	.2612817	.842007
sqft	.2900879	.0904748	3.21	0.003	.1073624	.4728134
age	-.7524605	1.097246	-0.69	0.502	-3.073675	1.568754
nfeatures	4.361055	13.59917	0.32	0.749	-22.67719	31.3993
ne	5.495913	34.97562	0.16	0.875	-63.95148	74.94331
custom	132.3453	43.26507	3.06	0.003	46.52087	218.1697
corner	-66.95606	40.55801	-1.65	0.102	-147.4264	13.51429
_cons	130.3491	64.03837	2.04	0.044	3.277868	257.4203

```
Transformations                            Average RVI      =     2.0039
                                           Complete DF      =        109
DF adjustment:    Small sample             DF:      min     =      23.04
                                                    avg     =      23.04
Within VCE type:            OLS                     max     =      23.04
       ratio: _b[age]/_b[sqft]
```

price	Coef.	Std. Err.	t	P>|t|	[95% Conf. Interval]	
ratio	-2.167045	3.460205	-0.63	0.537	-9.324358	4.990268

We use the nlcom syntax to specify the transformation: (ratio: _b[age]/_b[sqft]) defines the transformation and its name is ratio. All transformations must be specified following mi estimate and before the colon, and must be bound in parentheses.

A separate table containing the estimate of the ratio is displayed following the estimates of coefficients. If desired, we can suppress the table containing the estimates of coefficients by specifying the nocoef option. The header reports the average RVI due to nonresponse and the summaries of the degrees of freedom specific to the estimated transformations. Because we specified only one transformation, the minimum, average, and maximum degrees of freedom are the same. They correspond to the individual degrees of freedom for ratio.

See [MI] **mi estimate postestimation** for an example of linear transformation.

Potential problems that can arise when using mi estimate

There are two problems that can arise when using mi estimate:

1. The estimation sample varies across imputations.

2. Different covariates are omitted across the imputations.

mi estimate watches for and issues an error if either of these problems occur. Below we explain how each can arise and what to do about it. If you see one of these messages, be glad that mi estimate mentioned the problem, because otherwise, it might have gone undetected. A varying-estimation sample may result in biased or inefficient estimates. Different covariates being omitted always results in the combined results being biased.

If the first problem arises, mi estimate issues the error "estimation sample varies between $m = \#$ and $m = \#$". mi estimate expects that when it runs the estimation command on the first imputation, on the second, and so on, the estimation command will use the same observations in each imputation. mi estimate does not just count, it watches which observations are used.

Perhaps the difference is due to a past mistake, such as not having imputed all the missing values. Perhaps you even corrupted your mi data so that the imputed variable is missing in some imputations and not in others.

Another reason the error can arise is because you specified an if condition based on imputed or passive variables. mi estimate considers this a mistake but, if this is your intent, you can reissue the mi estimate command and include the esampvaryok option.

Finally, it is possible that the varying observations are merely a characteristic of the estimator when combined with the two different imputed datasets. In this case, just as in the previous one, you can reissue mi estimate with the esampvaryok option.

The easy way to diagnose why you got this error is to use mi xeq (see [MI] **mi xeq**) to run the estimation command separately on the two imputations mentioned in the error message. Alternatively, you can rerun the mi estimate command immediately with the esampvaryok option and with the esample(*varname*) option, which will create in new variable *varname* the e(sample) from each of the individual estimations. If you use the second approach, you must first mi convert your data to flong or flongsep if they are not recorded in that style already; see [MI] **mi convert** for details.

The second problem we mentioned concerns omitted variables as opposed to omitted observations. mi estimate reports "omitted variables vary" and goes on to mention the two imputations between which the variation was detected.

This can be caused when you include factor variables but did not specify base categories. It was the base categories that differed in the two imputations. That could happen if you specified i.group. By default, Stata chooses to omit the most frequent category. If group were imputed or passive, then the most frequent category could vary between two imputations. The solution is to specify the base category for yourself by typing, for instance, b2.group; see [U] **11.4.3 Factor variables**.

There are other possible causes. Varying omitted variables includes 1) different variables being omitted in the two imputations and 2) includes no variables being omitted in one imputation and, in the other, one or more variables being omitted.

When different variables are being omitted, it is usually caused by collinearity, and one of the variables needs to be dropped from the model. Variables x1 and x2 are collinear; sometimes the estimation command is choosing to omit x1 and other times, x2. The solution is that you choose which to omit by removing it from your model.

If no variables were omitted in one of the imputations, the problem is more difficult to explain. Say that you included i.group in your model, the base category remained the same for the two imputations, but in one of the imputations, no one is observed in group 3, and thus no coefficient for group 3 could be estimated. You choices are to accept that you cannot estimate a group 3 coefficient and combine group 3 with, say, group 4, or to drop all imputations in which there is no one in group 3. If you want to drop imputations 3, 9, and 12, you type mi set m -= (3,9,12); see [MI] **mi set**.

54   mi estimate — Estimation using multiple imputations

❑ Technical note

As we already mentioned, mi estimate obtains MI estimates by using the combination rules to pool results from the specified command executed separately on each imputation. As such, certain concepts (e.g., likelihood function) and most postestimation tools specific to the command may not be applicable to the MI estimates; see *Analysis of multiply imputed data* in [MI] **intro substantive**. MI estimates may not even have a valid variance–covariance matrix associated with them when the number of imputations is smaller than the number of estimated parameters. For these reasons, the system matrices e(b) and e(V) are not set by mi estimate. If desired, you can save the MI estimates and its variance–covariance estimate in e(b) and e(V) by specifying the post option. See [MI] **mi estimate postestimation** for postestimation tools available after mi estimate.

❑

Saved results

mi estimate saves the following in e():

Scalars
e(df_avg[_Q]_mi)	average degrees of freedom
e(df_c_mi)	complete degrees of freedom (if originally saved by *estimation_command* in e(df_r))
e(df_max[_Q]_mi)	maximum degrees of freedom
e(df_min[_Q]_mi)	minimum degrees of freedom
e(df_m_mi)	MI model test denominator (residual) degrees of freedom
e(df_r_mi)	MI model test numerator (model) degrees of freedom
e(esampvary_mi)	varying-estimation sample flag (0 or 1)
e(F_mi)	model test F statistic
e(k_exp_mi)	number of expressions (transformed coefficients)
e(M_mi)	number of imputations
e(N_mi)	number of observations (minimum, if varies)
e(N_min_mi)	minimum number of observations
e(N_max_mi)	maximum number of observations
e(p_mi)	MI model test p-value
e(rvi_avg[_Q]_mi)	average RVI
e(rvi_avg_F_mi)	average RVI associated with the residual degrees of freedom for model test
e(ufmi_mi)	1 if unrestricted FMI model test is performed, 0 if equal FMI model test is performed

Macros
    e(mi)   mi
    e(cmdline_mi)   command as typed
    e(prefix_mi)   mi estimate
    e(cmd_mi)   name of *estimation_command*
    e(cmd)   mi estimate (equals e(cmd_mi) when post is used)
    e(title_mi)   "Multiple-imputation estimates"
    e(wvce_mi)   title used to label within-imputation variance in the table header
    e(modeltest_mi)   title used to label the model test in the table header
    e(dfadjust_mi)   title used to label the degrees-of-freedom adjustment in the table header
    e(expnames_mi)   names of expressions specified in *spec*
    e(exp#_mi)   expressions of the transformed coefficients specified in *spec*
    e(rc_mi)   return codes for each imputation
    e(m_mi)   specified imputation numbers
    e(m_est_mi)   imputation numbers used in the computation
    e(names_vvl_mi)   command-specific e() macro names that contents varied across imputations
    e(names_vvm_mi)   command-specific e() matrix names that values varied across imputations
                          (excluding b, V, and Cns)
    e(names_vvs_mi)   command-specific e() scalar names that values varied across imputations

Matrices
    e(b)   MI estimates of coefficients (equals e(b_mi), saved only if post is used)
    e(V)   variance–covariance matrix (equals e(V_mi), saved only if post is used)
    e(Cns)   constraint matrix (for constrained estimation only; equals e(Cns_mi),
                          saved only if post is used)
    e(b[_Q]_mi)   MI estimates of coefficients (or transformed coefficients)
    e(V[_Q]_mi)   variance–covariance matrix (total variance)
    e(Cns_mi)   constraint matrix (for constrained estimation only)
    e(W[_Q]_mi)   within-imputation variance matrix
    e(B[_Q]_mi)   between-imputation variance matrix
    e(re[_Q]_mi)   parameter-specific relative efficiencies
    e(rvi[_Q]_mi)   parameter-specific RVIs
    e(fmi[_Q]_mi)   parameter-specific FMIs
    e(df[_Q]_mi)   parameter-specific degrees of freedom
    e(pise[_Q]_mi)   parameter-specific percentages increase in standard errors
    e(*vs_names*_vs_mi)   values of command-specific e() scalar *vs_names* that varied across imputations

vs_names include (but are not restricted to) df_r, N, N_strata, N_psu, N_pop, N_sub, N_postrata, N_stdize, N_subpop, N_over, and converged.

If transformations are specified, the corresponding estimation results are saved with the _Q_mi suffix, as described above.

Command-specific e() results that remain constant across imputations are also saved. Command-specific results that vary from imputation to imputation are posted as missing, and their names are saved in the corresponding macros e(names_vvl_mi), e(names_vvm_mi), and e(names_vvs_mi). For some command-specific e() scalars (see *vs_names* above), their values from each imputation are saved in a corresponding matrix with the _vs_mi suffix.

Methods and formulas

Let $\mathbf{q}$ define a column vector of parameters of interest. For example, $\mathbf{q}$ may be a vector of coefficients (or functions of coefficients) from a regression model. Let $\{(\widehat{\mathbf{q}}_i, \widehat{\mathbf{U}}_i): i = 1, 2, \ldots, M\}$ be the completed-data estimates of $\mathbf{q}$ and the respective variance–covariance estimates from M imputed datasets.

The MI estimate of $\mathbf{q}$ is

$$\overline{\mathbf{q}}_M = \frac{1}{M} \sum_{i=1}^{M} \widehat{\mathbf{q}}_i$$

The variance–covariance estimate (VCE) of $\overline{\mathbf{q}}_M$ (total variance) is

$$\mathbf{T} = \overline{\mathbf{U}} + (1 + \frac{1}{M})\mathbf{B}$$

where $\overline{\mathbf{U}} = \sum_{i=1}^{M} \widehat{\mathbf{U}}_i / M$ is the within-imputation variance–covariance matrix and $\mathbf{B} = \sum_{i=1}^{M} (\mathbf{q}_i - \overline{\mathbf{q}}_M)(\mathbf{q}_i - \overline{\mathbf{q}}_M)' / (M - 1)$ is the between-imputation variance–covariance matrix.

Methods and formulas are presented under the following headings:

> *Univariate case*
> *Multivariate case*

Univariate case

Let Q, $\overline{Q}_M$, B, $\overline{U}$, and T correspond to the scalar analogues of the above formulas. Univariate inferences are based on the approximation

$$T^{-1/2}(Q - \overline{Q}_M)^2 \sim t_\nu \qquad (1)$$

where t_ν denotes a Student's t distribution with ν degrees of freedom, which depends on the number of imputations, M, and the increase in variance of estimates due to missing data. Under the large-sample assumption with respect to complete data, the degrees of freedom is

$$\nu_{\text{large}} = (M - 1)\left(1 + \frac{1}{r}\right)^2 \qquad (2)$$

where

$$r = \frac{(1 + M^{-1})B}{\overline{U}} \qquad (3)$$

is an RVI due to missing data. Under the small-sample assumption, the degrees of freedom is

$$\nu_{\text{small}} = \left(\frac{1}{\nu_{\text{large}}} + \frac{1}{\widehat{\nu}_{\text{obs}}}\right)^{-1} \qquad (4)$$

where $\widehat{\nu}_{\text{obs}} = \nu_c(\nu_c + 1)(1 - \gamma)/(\nu_c + 3)$, $\gamma = (1 + 1/M)B/T$, and ν_c are the complete degrees of freedom, the degrees of freedom used for inference when data are complete (Barnard and Rubin 1999).

The sample-sample adjustment (4) is applied to the degrees of freedom ν when the specified command saves the residual degrees of freedom in e(df_r). This number of degrees of freedom is used as the complete degrees of freedom, ν_c, in the computation. (If e(df_r) varies across imputations, the smallest is used in the computation, resulting in conservative inference.) If e(df_r) is not set by the specified command or if the nosmall option is specified, then (2) is used to compute the degrees of freedom, ν.

Parameter-specific significance levels, confidence intervals, and degrees of freedom as reported by mi estimate are computed using the formulas above.

The percentage of standard-error increase due to missing data, as reported by mi estimate, dftable, is computed as $\{(T/\overline{U})^{1/2} - 1\} \times 100\%$.

The FMIs due to missing data and relative efficiencies reported by mi estimate, vartable are computed as follows.

The fraction of information about Q missing due to nonresponse is

$$\lambda = \frac{r + 2/(\nu_{\text{large}} + 3)}{r + 1}$$

where the RVI, r, is defined in (3).

The relative (variance) efficiency of using M imputations versus the infinite number of imputations is $RE = (1 + \lambda/M)^{-1}$ (Rubin 1987, 114).

Also see Rubin (1987, 76–77) and Schafer (1997, 109–111) for details.

Multivariate case

The approximation (1) can be generalized to the multivariate case:

$$(\mathbf{q} - \overline{\mathbf{q}}_M)\mathbf{T}^{-1}(\mathbf{q} - \overline{\mathbf{q}}_M)'/k \sim F_{k,\nu} \qquad (5)$$

where $F_{k,\nu}$ denotes an F distribution with $k = \text{rank}(T)$ numerator degrees of freedom and ν denominator degrees of freedom defined as in (2), where the RVI, r, is replaced with the average RVI, r_{ave}:

$$r_{\text{ave}} = (1 + 1/M)\text{tr}(\mathbf{B}\overline{\mathbf{U}}^{-1})/k$$

The approximation (5) is inadequate with a small number of imputations because the between-imputation variance, $\mathbf{B}$, cannot be estimated reliably based on small M. Moreover, when M is smaller than the number of estimated parameters, $\mathbf{B}$ does not have a full rank. As such, the total variance, $\mathbf{T}$, may not be a valid variance–covariance matrix for $\overline{\mathbf{q}}_M$.

One solution is to assume that the between-imputation and within-imputation matrices are proportional, that is $B = \overline{\lambda} \times \overline{U}$ (Rubin 1987, 78). This assumption implies that FMIs of all estimated parameters are equal. Under this assumption, approximation (5) becomes

$$(1 + r_{\text{ave}})^{-1}(\mathbf{q} - \overline{\mathbf{q}}_M)\overline{\mathbf{U}}^{-1}(\mathbf{q} - \overline{\mathbf{q}}_M)'/k \sim F_{k,\nu_\star} \qquad (6)$$

where $k = \text{rank}(U)$ and $\nu_\star$ is computed as described in Li, Meng, Raghunathan, and Rubin (1991, 1067).

Also see Rubin (1987, 77–78) and Schafer (1997, 112–114) for details.

We refer to (6) as an equal FMI test and to (5) as the unrestricted FMI test. By default, `mi estimate` uses the approximation (6) for the model test. If the `ufmitest` option is specified, it uses the approximation (5) for the model test.

Similar to the univariate case, the degrees of freedom $\nu_\star$ and ν are adjusted for small samples when the command saves the completed-data residual degrees of freedom in `e(df_r)`.

In the small-sample case, the degrees of freedom $\nu_\star$ is computed as described in Reiter (2007) (in the rare case, when $k(M-1) \leq 4$, $\nu_\star = (k+1)\nu_1/2$, where ν_1 is the degrees of freedom from Barnard and Rubin [1999]). In the small-sample case, the degrees of freedom ν is computed as described in Barnard and Rubin (1999) (J. P. Reiter, 2009, pers. comm.).

References

Barnard, J., and D. B. Rubin. 1999. Small-sample degrees of freedom with multiple imputation. *Biometrika* 86: 948–955.

Li, K.-H., X.-L. Meng, T. E. Raghunathan, and D. B. Rubin. 1991. Significance levels from repeated p-values with multiply-imputed data. *Statistica Sinica* 1: 65–92.

Reiter, J. P. 2007. Small-sample degrees of freedom for multi-component significance tests with multiple imputation for missing data. *Biometrika* 94: 502–508.

Royston, P., J. B. Carlin, and I. R. White. 2009. Multiple imputation of missing values: New features for mim. *Stata Journal*. Forthcoming.

Rubin, D. B. 1987. *Multiple Imputation for Nonresponse in Surveys*. New York: Wiley.

Schafer, J. L. 1997. *Analysis of Incomplete Multivariate Data*. Boca Raton, FL: Chapman & Hall/CRC.

Also see

[MI] **mi estimate using** — Estimation using previously saved estimation results

[MI] **mi estimate postestimation** — Postestimation tools for mi estimate

[MI] **intro substantive** — Introduction to multiple-imputation analysis

[MI] **intro** — Introduction to mi

[MI] **Glossary**

Title

> **mi estimate using** — Estimation using previously saved estimation results

Syntax

Compute MI estimates of coefficients using previously saved estimation results

    mi <u>est</u>imate using *miestfile* [, *options*]

Compute MI estimates of transformed coefficients using previously saved estimation results

    mi <u>est</u>imate [*spec*] using *miestfile* [, *options*]

where *spec* may be one or more terms of the form ([*name*:] *exp*). *exp* is any function of the parameter estimates allowed by nlcom; see [R] **nlcom**.

miestfile.ster contains estimation results previously saved by mi estimate, saving(*miestfile*); see [MI] **mi estimate**.

options	description
Options	
<u>nim</u>putations(*#*)	specify number of imputations to use; default is to use all saved imputations
<u>im</u>putations(*numlist*)	specify which imputations to use
<u>est</u>imations(*numlist*)	specify which estimation results to use
<u>ufmitest</u>	perform unrestricted FMI model test
<u>nosmall</u>	do not apply small-sample adjustment to degrees of freedom
Tables	
[<u>no</u>]<u>citable</u>	suppress/display standard estimation table containing parameter-specific confidence intervals; default is citable
<u>dftable</u>	display degrees-of-freedom table; dftable implies nocitable
<u>vartable</u>	display variance information about estimates; vartable implies citable
table_options	control table output
display_options	control spacing and display of omitted variables and base and empty cells
Reporting	
<u>level</u>(*#*)	set confidence level; default is level(95)
dots	display dots as estimations performed
<u>noisily</u>	display any output from nlcom if transformations are specified
trace	trace nlcom if transformations are specified; implies noisily
replay	replay command-specific results from each individual estimation in *miestfile*.ster; implies noisily
<u>cmdlegend</u>	display the command legend

Advanced	
errorok	allow estimation even when nlcom errors out in some imputations; such imputations are discarded from the analysis
†coeflegend	display coefficients' legend instead of coefficient table; implies nocitable and cannot be specified with citable or dftable
†nowarning	suppress the warning about varying estimation samples
†noerrnotes	suppress error notes associated with failed estimation results in *miestfile*.ster
†showimputations	show imputations saved in *miestfile*.ster
†*eform_option*	display coefficient table in exponentiated form
†post	post estimated coefficients and VCE to e(b) and e(V)

† coeflegend, nowarning, noerrnotes, showimputations, *eform_option*, and post do not appear in the dialog box.

table_options	description
noheader	suppress table header(s)
notable	suppress table(s)
nocoef	suppress table output related to coefficients
nocmdlegend	suppress command legend that appears in the presence of transformed coefficients when nocoef is used
notrcoef	suppress table output related to transformed coefficients
nolegend	suppress table legend(s)
nocnsreport	do not display constraints

See [MI] **mi estimate postestimation** for features available after estimation. To replay results, type mi estimate without arguments.

Menu

Statistics > Multiple imputation

Description

mi estimate using *miestfile* is for use after mi estimate: ..., saving(*miestfile*). It allows obtaining multiple-imputation (MI) estimates, including standard errors and confidence intervals, for transformed coefficients or the original coefficients, this time calculated on a subset of the imputations. The transformation can be linear or nonlinear.

Options

> Options

nimputations(#) specifies that the first # imputations be used; # must be $2 \leq \# \leq M$. The default is to use all imputations, M. Only one of nimputations(), imputations(), or estimations() may be specified.

imputations(*numlist*) specifies which imputations to use. The default is to use all of them. *numlist* must contain at least two numbers corresponding to the imputations saved in *miestfile*.ster. You can use the showimputations option to display imputations currently saved in *miestfile*.ster. Only one of nimputations(), imputations(), or estimations() may be specified.

estimations(*numlist*) does the same thing as imputations(*numlist*), but this time the imputations are numbered differently. Say that *miestfile*.ster was created by mi estimate and mi estimate was told to limit itself to imputations 1, 3, 5, and 9. With imputations(), the imputations are still numbered 1, 3, 5, and 9. With estimations(), they are numbered 1, 2, 3, and 4. Usually, one does not specify a subset of imputations when using mi estimate, and so usually, the imputations() and estimations() options are identical. The specified *numlist* must contain at least two numbers. Only one of nimputations(), imputations(), or estimations() may be specified.

ufmitest specifies that the unrestricted FMI (fraction missing information) model test be used. The default test performed assumes equal fractions of information missing due to nonresponse for all coefficients. This is equivalent to the assumption that the between-imputation and within-imputation variances are proportional. The unrestricted test may be preferable when this assumption is suspect provided that the number of imputations is large relative to the number of estimated coefficients.

nosmall specifies that no small-sample adjustment be made to the degrees of freedom. By default, individual tests of coefficients (and transformed coefficients) use the small-sample adjustment of Barnard and Rubin (1999), and the overall model test uses the small-sample adjustment of Reiter (2007).

Tables

All table options below may be specified at estimation time or when redisplaying previously estimated results.

citable and nocitable specify whether the standard estimation table containing parameter-specific confidence intervals be displayed. The default is citable. nocitable can be used with vartable to suppress the confidence-interval table.

dftable displays a table containing parameter-specific degrees of freedom and percentages of increase in standard errors due to nonresponse. dftable implies nocitable.

vartable displays a table reporting variance information about MI estimates. The table contains estimates of within-imputation variances, between-imputation variances, total variances, relative increases in variance due to nonresponse, fractions of information about parameter estimates missing due to nonresponse, and relative efficiencies for using finite M rather than a hypothetically infinite number of imputations. vartable implies citable.

table_options control the appearance of all displayed table output:

noheader suppresses all header information from the output. The table output is still displayed.

notable suppresses all tables from the output. The header information is still displayed.

nocoef suppresses the display of tables containing coefficient estimates. This option affects the table output produced by citable, dftable, and vartable.

nocmdlegend suppresses the table legend showing the command line, used to produce results in *miestfile*.ster, from the output. This legend appears above the tables containing transformed coefficients (or above the variance-information table if vartable is used) when nocoef is specified.

notrcoef suppresses the display of tables containing estimates of transformed coefficients (if specified). This option affects the table output produced by citable, dftable, and vartable.

nolegend suppresses all table legends from the output.

nocnsreport; see [R] **estimation options**.

display_options: <u>noomit</u>ted, vsquish, <u>noempty</u>cells, <u>base</u>levels, <u>allbase</u>levels; see [R] **estimation options**.

Reporting

level(*#*); see [R] **estimation options**.

dots specifies that dots be displayed as estimations of transformed coefficients are successfully completed. An **x** is displayed if nlcom fails to estimate one of the transformed coefficients specified in *spec*. This option is relevant only if transformations are specified.

noisily specifies that any output from nlcom, used to obtain the estimates of transformed coefficients, be displayed. This option is relevant only if transformations are specified.

trace traces the execution of nlcom. trace implies noisily and is relevant only if transformations are specified.

replay replays estimation results from *miestfile*.ster, previously saved by mi estimate, saving(*miestfile*). This option implies noisily.

cmdlegend requests that the command line corresponding to the estimation command used to produce the estimation results saved in *miestfile*.ster be displayed. cmdlegend may be specified at run time or when redisplaying results.

Advanced

errorok specifies that estimations of transformed coefficients that fail be skipped and the combined results be based on the successful estimation results. The default is that mi estimate stops if an individual estimation fails. If the *miestfile*.ster file contains failed estimation results, mi estimate using does not error out; it issues notes about which estimation results failed and discards these estimation results in the computation. You can use the noerrnotes option to suppress the display of the notes.

The following options are available with mi estimate using but are not shown in the dialog box:

coeflegend; see [R] **estimation options**. coeflegend implies nocitable and cannot be combined with citable or dftable.

nowarning suppresses the warning message at the bottom of table output that occurs if the estimation sample varies and esampvaryok is specified. See *Potential problems that can arise when using mi estimate* in [MI] **mi estimate** for details.

noerrnotes suppresses notes about failed estimation results. These notes appear when *miestfile*.ster contains estimation results, previously saved by mi estimate, saving(*miestfile*), from imputations for which the estimation command used with mi estimate failed to estimate parameters.

showimputations displays imputation numbers corresponding to the estimation results saved in *miestfile*.ster. showimputations may be specified at run time or when redisplaying results.

eform_option; see [R] ***eform_option***. mi estimate using reports results in the coefficient metric under which the combination rules are applied. You may use the appropriate *eform_option* to redisplay results in exponentiated form, if desired. If dftable is also specified, the reported degrees of freedom and percentage increases in standard errors are not adjusted and correspond to the original coefficient metric.

post requests that MI estimates of coefficients and their respective VCEs be posted in the usual way. This allows the use of *estimation_command*-specific postestimation tools with MI estimates. There are issues; see *Using the command-specific postestimation tools* in [MI] **mi estimate postestimation**. post may be specified at estimation time or when redisplaying previously estimated results.

Remarks

mi estimate using is convenient when refitting models using mi estimate would be tedious or time consuming. In such cases, you can perform estimation once and save the uncombined, individual results by specifying mi estimate's saving(*miestfile*) option. After that, you can repeatedly use mi estimate using *miestfile* to estimate linear and nonlinear transformations of coefficients or to obtain MI estimates using a subset of saved imputations.

mi estimate using performs the pooling step of the MI procedure; see [MI] **intro substantive**. That is, it combines completed-data estimates from the *miestfile*.ster file by applying Rubin's combination rules (Rubin 1987, 77).

▷ Example 1

Recall the analysis of house resale prices from *Example 2: Completed-data regression analysis* in [MI] **mi estimate**:

```
. use http://www.stata-press.com/data/r11/mhouses1993s30
(Albuquerque Home Prices Feb15-Apr30, 1993)

. mi estimate, saving(miest): regress price tax sqft age nfeatures ne custom corner

Multiple-imputation estimates              Imputations      =         30
Linear regression                          Number of obs    =        117
                                           Average RVI      =     0.5415
                                           Complete DF      =        109
DF adjustment:   Small sample              DF:      min     =      16.42
                                                    avg     =      72.83
                                                    max     =     101.18
Model F test:       Equal FMI              F(   7,   96.3)  =      45.63
Within VCE type:          OLS              Prob > F         =     0.0000

------------------------------------------------------------------------------
       price |      Coef.   Std. Err.      t    P>|t|     [95% Conf. Interval]
-------------+----------------------------------------------------------------
         tax |   .5516444   .1443319     3.82   0.000     .2612817    .842007
        sqft |   .2900879   .0904748     3.21   0.003     .1073624   .4728134
         age |  -.7524605   1.097246    -0.69   0.502    -3.073675   1.568754
    nfeatures|   4.361055   13.59917     0.32   0.749    -22.67719    31.3993
          ne |   5.495913   34.97562     0.16   0.875    -63.95148   74.94331
      custom |   132.3453   43.26507     3.06   0.003     46.52087   218.1697
      corner |  -66.95606   40.55801    -1.65   0.102    -147.4264   13.51429
       _cons |   130.3491   64.03837     2.04   0.044     3.277868   257.4203
------------------------------------------------------------------------------
```

In the above, we use the saving() option to save the individual completed-data estimates from a regression analysis in Stata estimation file miest.ster. We can now use mi estimate using to recombine the first 5 imputations, and ignoring the remaining 25, without reestimation:

```
. mi estimate using miest, ni(5)
```

Multiple-imputation estimates				Imputations	=	5
Linear regression				Number of obs	=	117
				Average RVI	=	0.2939
				Complete DF	=	109
DF adjustment:	Small sample			DF: min	=	7.51
				avg	=	61.46
				max	=	94.51
Model F test:	Equal FMI			F(7, 79.3)	=	53.00
Within VCE type:	OLS			Prob > F	=	0.0000

price	Coef.	Std. Err.	t	P>|t|	[95% Conf. Interval]	
tax	.5375429	.1399677	3.84	0.001	.2491601	.8259257
sqft	.2940729	.0890706	3.30	0.004	.1066042	.4815415
age	-.7325252	.7605651	-0.96	0.365	-2.506304	1.041254
nfeatures	3.696554	13.69565	0.27	0.788	-23.62366	31.01677
ne	8.303979	34.64444	0.24	0.811	-60.47857	77.08653
custom	129.2014	44.29287	2.92	0.004	41.25603	217.1468
corner	-63.77848	41.48302	-1.54	0.128	-146.1749	18.61793
_cons	135.4265	64.57713	2.10	0.039	7.184164	263.6688

We obtain results identical to those shown in the example in [MI] **mi estimate**.

We can also obtain estimates of transformed coefficients without refitting the models to the imputed dataset. Recall the example from *Example 4: Estimating transformations* in [MI] **mi estimate**, where we estimated the ratio of the coefficients for age and sqft. We can obtain the same results by using the following:

```
. mi estimate (ratio: _b[age]/_b[sqft]) using miest
```

Multiple-imputation estimates				Imputations	=	30
Linear regression				Number of obs	=	117
				Average RVI	=	0.5415
				Complete DF	=	109
DF adjustment:	Small sample			DF: min	=	16.42
				avg	=	72.83
				max	=	101.18
Model F test:	Equal FMI			F(7, 96.3)	=	45.63
Within VCE type:	OLS			Prob > F	=	0.0000

price	Coef.	Std. Err.	t	P>|t|	[95% Conf. Interval]	
tax	.5516444	.1443319	3.82	0.000	.2612817	.842007
sqft	.2900879	.0904748	3.21	0.003	.1073624	.4728134
age	-.7524605	1.097246	-0.69	0.502	-3.073675	1.568754
nfeatures	4.361055	13.59917	0.32	0.749	-22.67719	31.3993
ne	5.495913	34.97562	0.16	0.875	-63.95148	74.94331
custom	132.3453	43.26507	3.06	0.003	46.52087	218.1697
corner	-66.95606	40.55801	-1.65	0.102	-147.4264	13.51429
_cons	130.3491	64.03837	2.04	0.044	3.277868	257.4203

```
Transformations                              Average RVI   =     2.0039
                                             Complete DF   =        109
DF adjustment:      Small sample             DF:     min   =      23.04
                                                     avg   =      23.04
Within VCE type:           OLS                       max   =      23.04
        ratio: _b[age]/_b[sqft]
```

price	Coef.	Std. Err.	t	P>|t|	[95% Conf. Interval]	
ratio	-2.167045	3.460205	-0.63	0.537	-9.324358	4.990268

The results are the same as in the example in [MI] **mi estimate**.

◁

For more examples, see [MI] **mi estimate postestimation**.

Saved results

See *Saved results* in [MI] **mi estimate**.

Methods and formulas

See *Methods and formulas* in [MI] **mi estimate**.

References

Barnard, J., and D. B. Rubin. 1999. Small-sample degrees of freedom with multiple imputation. *Biometrika* 86: 948–955.

Li, K.-H., X.-L. Meng, T. E. Raghunathan, and D. B. Rubin. 1991. Significance levels from repeated p-values with multiply-imputed data. *Statistica Sinica* 1: 65–92.

Reiter, J. P. 2007. Small-sample degrees of freedom for multi-component significance tests with multiple imputation for missing data. *Biometrika* 94: 502–508.

Rubin, D. B. 1987. *Multiple Imputation for Nonresponse in Surveys*. New York: Wiley.

Also see

[MI] **mi estimate** — Estimation using multiple imputations

[MI] **mi estimate postestimation** — Postestimation tools for mi estimate

[MI] **intro substantive** — Introduction to multiple-imputation analysis

[MI] **intro** — Introduction to mi

[MI] **Glossary**

Title

> **mi estimate postestimation** — Postestimation tools for mi estimate

Syntax

Test that coefficients are zero

    mi test *coeflist*

Test that coefficients within a single equation are zero

    mi test [*eqno*] [: *coeflist*]

Test that subsets of coefficients are zero (full syntax)

    mi test (*spec*) [(*spec*) ...] [, <u>ufmi</u>test nosmall <u>cons</u>tant]

Test that subsets of transformed coefficients are zero

    mi <u>testtr</u>ansform *name* [(*name*) ...] [, <u>ufmi</u>test nosmall <u>noleg</u>end]

options	description
Test	
<u>ufmi</u>test	perform unrestricted FMI model test
nosmall	do not apply small-sample correction to degrees of freedom
<u>cons</u>tant	include the constant in coefficients to be tested
<u>noleg</u>end	suppress legend about transformed coefficients

coeflist may contain factor variables and time-series operators; see [U] **11.4.3 Factor variables** and [U] **11.4.4 Time-series varlists**.

spec is
    coeflist
    [*eqno*] [: *coeflist*]

coeflist is
    coef [*coef* ...]
    [*eqno*]*coef* [[*eqno*]*coef*...]
    [*eqno*]_b[*coef*] [[*eqno*]_b[*coef*]...]

eqno is
    ##
    eqname

coef identifies a coefficient in the model; see [R] **test** for details. *eqname* is an equation name.

name is an expression name as specified with mi estimate or mi estimate using (see [MI] **mi estimate** or [MI] **mi estimate using**).

Menu

Statistics > Multiple imputation

Description

mi test performs joint tests of coefficients.

mi testtransform performs joint tests of transformed coefficients as specified with mi estimate or mi estimate using (see [MI] **mi estimate** or [MI] **mi estimate using**).

Options

⌐ Test ┐

ufmitest specifies that the unrestricted fraction missing information (FMI) model test be used. The default test performed assumes equal fractions of information missing due to nonresponse for all coefficients. This is equivalent to the assumption that the between-imputation and within-imputation variances are proportional. The unrestricted test may be preferable when this assumption is suspect provided that the number of imputations is large relative to the number of estimated coefficients.

nosmall specifies that no small-sample adjustment be made to the degrees of freedom. By default, individual tests of coefficients (and transformed coefficients) use the small-sample adjustment of Barnard and Rubin (1999), and the overall model test uses the small-sample adjustment of Reiter (2007).

constant specifies that _cons be included in the list of coefficients to be tested when using the [eqno] form of spec. The default is to not include _cons.

nolegend suppresses the legend about transformed coefficients from the output.

Remarks

Remarks are presented under the following headings:

>Introduction
>Overview
>Example 1: Testing subsets of coefficients equal to zero
>Example 2: Testing linear hypotheses
>Example 3: Testing nonlinear hypotheses
>Using the command-specific postestimation tools

Introduction

After estimation by mi estimate: *estimation_command*, you may not use the standard postestimation commands such as test, testnl, or predict; nor may you use *estimation_command*-specific postestimation commands such as estat. As we have mentioned often, mi estimate is its own estimation command, and the postestimation commands available after mi estimate (and mi estimate using) are mi test and mi testtransform.

The major issue arising when performing tests after MI estimation is the validity of the variance–covariance estimator (VCE) of the MI estimates. MI variance consists of the two sources of variation: within-imputation variation and between-imputation variation. With a small number of imputations, the estimate of the between-imputation variance–covariance matrix is imprecise. In fact, when the number of imputations is less than or equal to the number of the estimated parameters, the between-imputation matrix does not even have a full rank. As such, the estimated VCE may not be a valid variance–covariance matrix and thus not suitable for joint inference.

One solution to this problem was proposed by Rubin (1987) and Li et al. (1991). The idea is to assume that the between-imputation variance is proportional to the within-imputation variance. This assumption implies equal FMIs for all jointly tested parameters. Li et al. (1991) found that the procedure performs well in terms of power and maintaining the significance level even with moderately variable FMIs. `mi test` and `mi testtransform`, by default, perform tests using this procedure.

When the number of imputations is large enough relative to the number of tested parameters so that the corresponding VCE is trustworthy, you can request the unrestricted FMI test by specifying the `ufmitest` option. The unrestricted FMI test is the conventional test described by Rubin (1987, 77).

For testing nonlinear hypotheses, direct application of the conventional delta method to the estimated coefficients may not be feasible when the number of imputations is small enough that the VCE of the MI estimates cannot be used for inference. To test these hypotheses, one can first obtain MI estimates of the transformed coefficients by applying Rubin's combination rules to the transformed completed-data estimates, and then apply the above MI-specific hypotheses tests to the combined transformed estimates. The first step can be done by specifying expressions with `mi estimate` (or `mi estimate using`). The second step is performed with `mi testtransform`. `mi testtransform` uses the same method to test transformed coefficients as `mi test` uses to test coefficients.

Overview

Use `mi test` to perform joint tests that coefficients are equal to zero:

```
. mi estimate: y x1 x2 x3 x4
. mi test x2 x3 x4
```

Use `mi testtransform`, however, to perform tests of more general linear hypotheses, such as `_b[x1]=_b[x2]`, or `_b[x1]=_b[x2]` and `_b[x1]=_b[x3]`. Testing general linear hypotheses requires estimation of between and within variances corresponding to the specific hypotheses and requires recombining the imputation-specific estimation results. One way you could do that would be to reestimate the model and include the additional parameters during the estimation step. To test `_b[x1]=_b[x2]`, you could type

```
. mi estimate (diff: _b[x1]-_b[x2]): regress y x1 x3 x3 x4
. mi testtransform diff
```

A better approach, however, is to save each of the imputation-specific results at the time the original model is fit and then later recombine results using `mi estimate using`. To save the imputation-specific results, specify `mi estimate`'s `saving()` option when the model is originally fit:

```
. mi estimate, saving(myresults): regress y x1 x2 x3 x4
```

To test `_b[x1]=_b[x2]`, you type

```
. mi estimate (diff: _b[x1]-_b[x2]) using myresults
. mi testtransform diff
```

The advantage of this approach is that you can test additional hypotheses without refitting the model. For instance, if we now wanted to test `_b[x1]=_b[x2]` and `_b[x1]=_b[x3]`, we could type

```
. mi estimate (diff1: _b[x1]-_b[x2]) (diff2: _b[x1]=_b[x3]) using myresults
. mi testtransform diff1 diff2
```

To test nonlinear hypotheses, such as _b[x1]/_b[x2]=_b[x3]/_b[x4], we could then type

```
. mi estimate (diff: _b[x1]/_b[x2]-_b[x3]/_b[x4]) using myresults
. mi testtransform diff
```

Example 1: Testing subsets of coefficients equal to zero

We are going to test that tax, sqft, age, nfeatures, ne, custom, and corner are in the regression analysis of house resale prices we performed in *Example 1: Completed-data logistic analysis* of [MI] **mi estimate**. Following the advice above, when we fit the model, we are going to save the imputation-specific results even though we will not need them in this example; we will need them in the following examples.

```
. use http://www.stata-press.com/data/r11/mhouses1993s30
(Albuquerque Home Prices Feb15-Apr30, 1993)
. mi estimate, saving(miest): regress price tax sqft age nfeatures ne custom corner
Multiple-imputation estimates                   Imputations       =         30
Linear regression                               Number of obs     =        117
                                                Average RVI       =     0.5415
                                                Complete DF       =        109
DF adjustment:   Small sample                   DF:     min       =      16.42
                                                        avg       =      72.83
                                                        max       =     101.18
Model F test:       Equal FMI                   F(   7,   96.3)   =      45.63
Within VCE type:          OLS                   Prob > F          =     0.0000
```

price	Coef.	Std. Err.	t	P>|t|	[95% Conf. Interval]
tax	.5516444	.1443319	3.82	0.000	.2612817	.842007
sqft	.2900879	.0904748	3.21	0.003	.1073624	.4728134
age	-.7524605	1.097246	-0.69	0.502	-3.073675	1.568754
nfeatures	4.361055	13.59917	0.32	0.749	-22.67719	31.3993
ne	5.495913	34.97562	0.16	0.875	-63.95148	74.94331
custom	132.3453	43.26507	3.06	0.003	46.52087	218.1697
corner	-66.95606	40.55801	-1.65	0.102	-147.4264	13.51429
_cons	130.3491	64.03837	2.04	0.044	3.277868	257.4203

In the above, we use the saving() option to create a Stata estimation file, miest.ster, containing imputation-specific estimation results.

mi estimate reports the joint test of all coefficients equal to zero in the header. We can reproduce this test with mi test by typing

```
. mi test tax sqft age nfeatures ne custom corner
note: assuming equal fractions of missing information
 ( 1)  tax = 0
 ( 2)  sqft = 0
 ( 3)  age = 0
 ( 4)  nfeatures = 0
 ( 5)  ne = 0
 ( 6)  custom = 0
 ( 7)  corner = 0
         F(   7,   96.3) =     45.63
              Prob > F   =    0.0000
```

and obtain results identical to those from mi estimate.

We can test that a subset of coefficients, say, sqft and tax, are equal to zero by typing

```
. mi test sqft tax
note: assuming equal fractions of missing information

 ( 1)  sqft = 0
 ( 2)  tax = 0

       F(  2,  85.7) =    88.42
            Prob > F =   0.0000
```

Example 2: Testing linear hypotheses

Now we want to test the equality of the coefficients for sqft and tax. Following our earlier suggestion, we use mi estimate using to estimate the difference between coefficients (and avoid refitting the models) and then use mi testtransform to test that the difference is zero:

```
. mi estimate (diff: _b[tax]-_b[sqft]) using miest, nocoef

Multiple-imputation estimates                   Imputations       =         30
Linear regression                               Number of obs     =        117
                                                Average RVI       =     0.7933
                                                Complete DF       =        109
DF adjustment:    Small sample                  DF:      min      =      42.55
                                                         avg      =      42.55
Within VCE type:            OLS                          max      =      42.55

      command:  regress price tax sqft age nfeatures ne custom corner
         diff:  _b[tax]-_b[sqft]
```

price	Coef.	Std. Err.	t	P>|t|	[95% Conf. Interval]
diff	.2615564	.228952	1.14	0.260	-.200309 .7234219

```
. mi testtransform diff
note: assuming equal fractions of missing information

        diff:  _b[tax]-_b[sqft]

 ( 1)  diff = 0

       F(  1,  42.6) =    1.31
            Prob > F =   0.2597
```

We suppress the display of the coefficient table by specifying the nocoef option with mi estimate using. Notice that we obtain the same results from the F test as those of the t test reported in the transformation table.

Similarly, we can test whether three coefficients are jointly equal:

```
. mi estimate (diff1: _b[tax]-_b[sqft]) (diff2: _b[custom]-_b[tax]) using miest,
> nocoef
Multiple-imputation estimates              Imputations     =         30
Linear regression                          Number of obs   =        117
                                           Average RVI     =     0.4198
                                           Complete DF     =        109
DF adjustment:    Small sample             DF:      min    =      42.55
                                                    avg    =      71.88
Within VCE type:          OLS                       max    =     101.21

       command:  regress price tax sqft age nfeatures ne custom corner
         diff1:  _b[tax]-_b[sqft]
         diff2:  _b[custom]-_b[tax]
```

price	Coef.	Std. Err.	t	P>|t|	[95% Conf. Interval]	
diff1	.2615564	.228952	1.14	0.260	-.200309	.7234219
diff2	131.7936	43.2603	3.05	0.003	45.97905	217.6082

```
. mi testtr diff1 diff2
note: assuming equal fractions of missing information
         diff1:  _b[tax]-_b[sqft]
         diff2:  _b[custom]-_b[tax]
 ( 1)  diff1 = 0
 ( 2)  diff2 = 0
        F(  2,   87.9) =     3.98
             Prob > F =    0.0221
```

We estimate two differences, _b[tax]-_b[sqft] and _b[custom]-_b[tax], using mi estimate using and test whether they are jointly equal to zero by using mi testtransform.

We can perform tests of other hypotheses similarly by reformulating the hypotheses of interest such that we are testing equality to zero.

Example 3: Testing nonlinear hypotheses

In the examples above, we tested linear hypotheses. Testing nonlinear hypotheses is no different. We simply replace the specification of linear expressions in mi estimate using with the nonlinear expressions corresponding to the tests of interest.

For example, let's test that the ratio of the coefficients for tax and sqft is one, an equivalent but less efficient way of testing whether the two coefficients are the same. Similarly to the earlier example, we specify the corresponding nonlinear expression with mi estimate using and then use mi testtransform to test that the ratio is one:

(Continued on next page)

```
. mi estimate (rdiff: _b[tax]/_b[sqft] - 1) using miest, nocoef
Multiple-imputation estimates              Imputations     =         30
Linear regression                          Number of obs   =        117
                                           Average RVI     =     0.6833
                                           Complete DF     =        109
DF adjustment:    Small sample             DF:     min     =      46.72
                                                   avg     =      46.72
Within VCE type:          OLS                      max     =      46.72
        command: regress price tax sqft age nfeatures ne custom corner
        rdiff:   _b[tax]/_b[sqft] - 1
```

price	Coef.	Std. Err.	t	P>|t|	[95% Conf. Interval]
rdiff	1.051733	1.228478	0.86	0.396	-1.420042 3.523508

```
. mi testtr rdiff
note: assuming equal fractions of missing information
        rdiff: _b[tax]/_b[sqft] - 1
 ( 1)  rdiff = 0
        F(  1,  46.7) =    0.73
             Prob > F =    0.3963
```

Notice that we do not need to use mi testtransform (or mi test) to test one transformation (or coefficient) because the corresponding test is provided in the output from mi estimate using.

Using the command-specific postestimation tools

After mi estimate: *estimation_command*, you may not use *estimation_command*'s postestimation features. More correctly, you may not use them unless you specify mi estimate's post option:

    . mi estimate, post: *estimation_command* ...

Specifying post causes many statistical issues, so do not be casual about specifying it.

First, the MI estimate of the VCE is poor unless the number of imputations, M, is sufficiently large. How large is uncertain, but you should not be thinking $M = 20$ rather than $M = 5$; you should be thinking of M in the hundreds. What is statistically true is that, asymptotically in M (and in the number of observations, N), the MI estimated coefficients approach normality and the VCE becomes well estimated.

Second, there are substantive issues about what is meant by prediction after MI estimation that you are going to have to resolve for yourself. There is no one estimation sample. There are M of them, and as we have just argued, M is large. Do not expect postestimation commands that depend on predicted values such as margins, lroc, and the like, to produce correct results, if they produce results at all.

Which brings us to the third point. Even when you specify mi estimate's post option, mi estimate still does not post everything the estimation command expects to see. It does not post likelihood values, for instance, because there is no counterpart after MI estimation. Thus, you should be prepared to see unexpected and inelegant error messages if you use a postestimation command that depends on an unestimated and unposted result.

All of which is to say that if you specify the post option, you have a responsibility beyond the usual to ensure the validity of any statistical results.

Saved results

`mi test` and `mi testtransform` save the following in `r()`:

Scalars
    `r(p)`   two-sided p-value
    `r(F)` F statistic
    `r(df)`   test constraints degrees of freedom
    `r(df_r)`   residual degrees of freedom
    `r(drop)`   1 if constraints were dropped, 0 otherwise
    `r(dropped_i)` index of ith constraint dropped

Methods and formulas

`mi test` and `mi testtransform` use the methodology described in *Multivariate case* under *Methods and formulas* of [MI] **mi estimate**, where we replace $\mathbf{q}$ with $\mathbf{Rq} - \mathbf{r}$ and $\mathbf{q}_0 = \mathbf{0}$ for the test H_0: $\mathbf{Rq} = \mathbf{r}$.

References

Barnard, J., and D. B. Rubin. 1999. Small-sample degrees of freedom with multiple imputation. *Biometrika* 86: 948–955.

Li, K.-H., X.-L. Meng, T. E. Raghunathan, and D. B. Rubin. 1991. Significance levels from repeated p-values with multiply-imputed data. *Statistica Sinica* 1: 65–92.

Reiter, J. P. 2007. Small-sample degrees of freedom for multi-component significance tests with multiple imputation for missing data. *Biometrika* 94: 502–508.

Rubin, D. B. 1987. *Multiple Imputation for Nonresponse in Surveys*. New York: Wiley.

Also see

[MI] **mi estimate** — Estimation using multiple imputations

[MI] **mi estimate using** — Estimation using previously saved estimation results

[MI] **intro substantive** — Introduction to multiple-imputation analysis

[MI] **intro** — Introduction to mi

[MI] **Glossary**

Title

mi expand — Expand mi data

Syntax

mi expand [=]*exp* [*if*] [, *options*]

options	description
generate(*newvar*)	create *newvar*; 0 = original, 1 = expanded
noupdate	see [MI] **noupdate option**

Menu

Statistics > Multiple imputation

Description

mi expand is expand (see [D] **expand**) for mi data. The syntax is identical to expand except that in *range* is not allowed and the noupdate option is allowed.

mi expand replaces each observation in the dataset with n copies of the observation, where n is equal to the required expression rounded to the nearest integer. If the expression is less than 1 or equal to missing, it is interpreted as if it were 1, meaning that the observation is retained but not duplicated.

Options

generate(*newvar*) creates new variable *newvar* containing 0 if the observation originally appeared in the dataset and 1 if the observation is a duplication.

noupdate in some cases suppresses the automatic mi update this command might perform; see [MI] **noupdate option**.

Remarks

mi expand amounts to performing expand on $m = 0$, then duplicating the result on $m = 1$, $m = 2, \ldots, m = M$, and then combining the result back into mi format. Thus, if the requested expansion specified by *exp* is a function of an imputed, passive, varying, or super-varying variable, it is the values of the variable in $m = 0$ that will be used to produce the result for $m = 1$, $m = 2$, $\ldots, m = M$, too.

Also see

[MI] **intro** — Introduction to mi

[D] **expand** — Duplicate observations

Title

> **mi export** — Export mi data

Syntax

    mi export nhanes1 ...   (see [MI] **mi export nhanes1**)

    mi export ice ...   (see [MI] **mi export ice**)

Description

Use `mi export nhanes1` to export data in the format used by the National Health and Nutrition Examination Survey.

Use `mi export ice` to export data in the format used by `ice` (Royston 2004, 2005a, 2005b).

If and when other standards develop for recording multiple-imputation data, other `mi export` subcommands will be added.

Remarks

If you wish to send data to other Stata users, ignore `mi export` and just send them your mi dataset(s).

To send data to users of other packages, however, you will have to negotiate the format you will use. The easiest way to send data to non–Stata users is probably to `mi convert` (see [MI] **mi convert**) your data to flongsep and then use `outfile` (see [D] **outfile**), outsheet (see [D] **outsheet**), or a transfer program such as Stat/Transfer. Also see [U] **21 Inputting data**.

References

Royston, P. 2004. Multiple imputation of missing values. *Stata Journal* 4: 227–241.

———. 2005a. Multiple imputation of missing values: Update. *Stata Journal* 5: 188–201.

———. 2005b. Multiple imputation of missing values: Update of ice. *Stata Journal* 5: 527–536.

Also see

[MI] **intro** — Introduction to mi

[MI] **mi export nhanes1** — Export mi data to NHANES format

[MI] **mi export ice** — Export mi data to ice format

Title

> **mi export ice** — Export mi data to ice format

Syntax

> mi export ice [, clear]

Menu

Statistics > Multiple imputation

Description

> mi export ice converts the mi data in memory to ice format. See Royston (2004, 2005a, 2005b) for a description of ice.

Option

> clear specifies that it is okay to replace the data in memory even if they have changed since they were last saved to disk.

Remarks

> mi export ice is the inverse of mi import ice (see [MI] **mi import ice**). Below we use mi export ice to convert miproto.dta to ice format. miproto.dta happens to be in wide form, but that is irrelevant.

```
. use http://www.stata-press.com/data/r11/miproto
(mi prototype)
. mi describe
  Style: wide
         last mi update 13apr2009 12:40:54, 1 day ago
  Obs.:  complete          1
         incomplete        1  (M = 2 imputations)
         ─────────────────────
         total             2
  Vars.: imputed: 1; b(1)
         passive: 1; c(1)
         regular: 1; a
         system:  1; _mi_miss
         (there are no unregistered variables)
. list
```

	a	b	c	_1_b	_2_b	_1_c	_2_c	_mi_miss
1.	1	2	3	2	2	3	3	0
2.	4	.	.	4.5	5.5	8.5	9.5	1

```
. mi export ice
```

```
. list, separator(2)

     +---------------------------+
     |  a     b     c   _mj  _mi |
     |---------------------------|
  1. |  1     2     3     0    1 |
  2. |  4     .     .     0    2 |
     |---------------------------|
  3. |  1     2     3     1    1 |
  4. |  4   4.5   8.5     1    2 |
     |---------------------------|
  5. |  1     2     3     2    1 |
  6. |  4   5.5   9.5     2    2 |
     +---------------------------+
```

References

Royston, P. 2004. Multiple imputation of missing values. *Stata Journal* 4: 227–241.

———. 2005a. Multiple imputation of missing values: Update. *Stata Journal* 5: 188–201.

———. 2005b. Multiple imputation of missing values: Update of ice. *Stata Journal* 5: 527–536.

Also see

[MI] **intro** — Introduction to mi

[MI] **mi export** — Export mi data

[MI] **mi import ice** — Import ice-format data into mi

Title

> **mi export nhanes1** — Export mi data to NHANES format

Syntax

> mi export nhanes1 *filenamestub* [, *options odd_options*]

options	description
replace	okay to replace existing files
uppercase	uppercase prefix and suffix
<u>p</u>assiveok	include passive variables

odd_options	description
nacode(#)	not applicable code; default is 0
obscode(#)	observed code; default is 1
impcode(#)	imputed code; default is 2
<u>impp</u>refix("*string*" "*string*")	variable prefix; default is "" ""
<u>imps</u>uffix("*string*" "*string*")	variable suffix; default is "if" "mi"

Note: The *odd_options* are not specified unless you want to create results that are nhanes1-like but not really nhanes1 format.

Menu

Statistics > Multiple imputation

Description

mi export nhanes1 writes the mi data in memory to disk files in nhanes1 format. The files will be named *filenamestub*.dta, *filenamestub*1.dta, *filenamestub*2.dta, and so on. In addition to the variables in the original mi data, new variable seqn recording the sequence number will be added. After using mi export nhanes1, you can use outfile (see [D] **outfile**) or outsheet (see [D] **outsheet**) or a transfer program such as Stat/Transfer to convert the resulting .dta files into a format suitable for sending to a non-Stata user. Also see [U] **21 Inputting data**.

mi export nhanes1 leaves the data in memory unchanged.

Options

replace indicates that it is okay to overwrite existing files.

uppercase specifies that the new sequence variable SEQN and the variable suffixes IF and MI are to be in uppercase. The default is lowercase. (More correctly, when generalizing beyond nhanes1 format, the uppercase option specifies that SEQN be created in uppercase along with all prefixes and suffixes.)

passiveok specifies that passive variables are to be written as if they were imputed variables. The default is to issue an error if passive variables exist in the original data.

nacode(#), obscode(#), and impcode(#) are optional and are never specified when reading true nhanes1 data. The default nacode(0) obscode(1) impcode(2) corresponds to the nhanes1 definition. These options allow changing the codes for not applicable, observed, and imputed.

impprefix("*string*" "*string*") and impsuffix("*string*" "*string*") are optional and are never specified when reading true nhanes1 data. The default impprefix("" "") impsuffix("if" "mi") corresponds to the nhanes1 definition. These options allow setting different prefixes and suffixes.

Remarks

mi export nhanes1 is the inverse of mi import nhanes1; see [MI] **mi import nhanes1** for a description of the nhanes1 format.

Below we use mi export nhanes1 to convert miproto.dta to nhanes1 format. miproto.dta happens to be in wide form, but that is irrelevant.

```
. use http://www.stata-press.com/data/r11/miproto
(mi prototype)
. mi describe
  Style:  wide
          last mi update 13apr2009 12:40:54, 1 day ago
   Obs.:  complete        1
          incomplete      1   (M = 2 imputations)

          total           2
  Vars.:  imputed: 1; b(1)
          passive: 1; c(1)
          regular: 1; a
          system:  1; _mi_miss
          (there are no unregistered variables)
. list
```

	a	b	c	_1_b	_2_b	_1_c	_2_c	_mi_miss
1.	1	2	3	2	2	3	3	0
2.	4	.	.	4.5	5.5	8.5	9.5	1

```
. mi export nhanes1 mynh, passiveok replace
files mynh.dta mynh1.dta mynh2.dta created
. use mynh
(mi prototype)
. list
```

	seqn	a	bif	cif
1.	1	1	1	1
2.	2	4	2	2

```
. use mynh1
(mi prototype)
. list

     +---------------------+
     | seqn   a   bmi  cmi |
     |---------------------|
  1. |    1   1     2    3 |
  2. |    2   4   4.5  8.5 |
     +---------------------+

. use mynh2
(mi prototype)
. list

     +---------------------+
     | seqn   a   bmi  cmi |
     |---------------------|
  1. |    1   1     2    3 |
  2. |    2   4   5.5  9.5 |
     +---------------------+
```

Also see

[MI] **intro** — Introduction to mi

[MI] **mi export** — Export mi data

[MI] **mi import nhanes1** — Import NHANES-format data into mi

Title

> **mi extract** — Extract original or imputed data from mi data

Syntax

mi extract # [, *options*]

where $0 \leq \# \leq M$

options	description
clear	okay to replace unsaved data in memory
esample(...)	rarely specified option
esample(*varname*)	... syntax when $\# > 0$
esample(*varname* $\#_e$)	... syntax when $\# = 0$; $1 \leq \#_e \leq M$

Menu

Statistics > Multiple imputation

Description

mi extract # replaces the data in memory with the data for $m = \#$. The data are not mi set.

Options

clear specifies that it is okay to replace the data in memory even if the current data have not been saved to disk.

esample(*varname* [$\#_e$]) is rarely specified. It is for use after mi estimate (see [MI] **mi estimate**) when the esample(*newvar*) option was specified to save in *newvar* the e(sample) for $m = 1$, $m = 2$, ..., $m = M$. It is now desired to extract the data for one m and for e(sample) set correspondingly.

mi extract #, esample(*varname*), $\# > 0$, is the usual case in this unlikely event. One extracts one of the imputation datasets and redefines e(sample) based on the e(sample) previously saved for $m = \#$.

The odd case is mi extract 0, esample(*varname* $\#_e$), where $\#_e > 0$. One extracts the original data but defines e(sample) based on the e(sample) previously saved for $m = \#_e$.

Specifying the esample() option changes the sort order of the data.

Remarks

If you wanted to give up on `mi` and just get your original data back, you could type

```
. mi extract 0
```

You might do this if you wanted to send your original data to a coworker or you wanted to try a different approach to dealing with the missing values in these data. Whatever the reason, the result is that the original data replace the data in memory. The data are not `mi set`. Your original `mi` data remain unchanged.

If you suspected there was something odd about the imputations in $m = 3$, you could type

```
. mi extract 3
```

You would then have a dataset in memory that looked just like your original, except the missing values of the imputed and passive variables would be replaced with the imputed and passive values from $m = 3$. The data are not `mi set`. Your original data remain unchanged.

Also see

[MI] **intro** — Introduction to mi

Title

mi import — Import data into mi

Syntax

    mi import nhanes1 ...

    mi import ice ...

    mi import flong ...

    mi import flongsep ...

    mi import wide ...

See [MI] **mi import nhanes1**, [MI] **mi import ice**, [MI] **mi import flong**, [MI] **mi import flongsep**, and [MI] **mi import wide**.

Description

mi import imports into mi data that contain original data and imputed values.

Remarks

Remarks are presented under the following headings:

> When to use which mi import command
> Import data into Stata before importing into mi
> Using mi import nhanes1, ice, flong, and flongsep

When to use which mi import command

mi import nhanes1 imports data recorded in the format used by the National Health and Nutrition Examination Survey (NHANES) produced by the National Center for Health Statistics of the U.S. Centers for Disease Control and Prevention (CDC); see http://www.cdc.gov/nchs/nhanes.htm.

mi import ice imports data recorded in the format used by ice (Royston 2004, 2005a, 2005b).

mi import flong and mi import flongsep import data that are in flong- and flongsep-like format, which is to say, the data are repeated for $m = 0$, $m = 1$, ..., and $m = M$. mi import flong imports data in which the information is contained in one file. mi import flongsep imports data in which the information is recorded in a collection of files.

mi import wide imports data that are in wide-like format, where additional variables are used to record the imputed values.

Import data into Stata before importing into mi

With the exception of `mi import ice`, you must import the data into Stata before you can use `mi import` to import the data into `mi`. `mi import ice` is the exception only because the data are already in Stata format. That is, `mi import` requires that the data be stored in Stata-format `.dta` datasets. You perform the initial import into Stata by using `infile` (see [D] **infile (free format)** or [D] **infile (fixed format)**), `infix` (see [D] **infix (fixed format)**), `fdause` (see [D] **fdasave**), `odbc` (see [D] **odbc**), or a transfer program such as Stat/Transfer; also see [U] **21 Inputting data**.

Using mi import nhanes1, ice, flong, and flongsep

Import commands `mi import nhanes1` and `mi import flongsep` produce an flongsep result; `mi import ice` and `mi import flong` produce an flong result. You can use `mi convert` (see [MI] **mi convert**) afterward to convert the result to another style, and we usually recommend that. Before doing that, however, you need to examine the freshly imported data and verify that all imputed and passive variables are registered correctly. If they are not registered correctly, you risk losing imputed values.

To perform this verification, use the `mi describe` (see [MI] **mi describe**) and `mi varying` (see [MI] **mi varying**) commands immediately after `mi import`:

    . mi import ...
    . mi describe
    . mi varying

`mi describe` will list the registration status of the variables. `mi varying` will report the varying and super-varying variables. Verify that all varying variables are registered as imputed or passive. If one or more is not, register them now:

    . mi register imputed *forgottenvar*
    . mi register passive *another_forgottenvar*

There is no statistical distinction between imputed and passive variables, so you may register variables about which you are unsure either way. If an unregistered variable is found to be varying and you are convinced that is an error, register the variable as regular:

    . mi register regular *variable_in_error*

Next, if `mi varying` reports that your data contain any super-varying variables, determine whether the variables are due to errors in the source data or really are intended to be super varying. If they are errors, register the variables as imputed, passive, or regular, as appropriate. Leave any intended super-varying variables unregistered, however, and make a note to yourself: never convert these data to the wide or mlong styles. Data with super-varying variables can be stored only in the flong and flongsep styles.

Now run `mi describe` and `mi varying` again:

    . mi describe
    . mi varying

Ensure that you have registered variables correctly, and, if necessary, repeat the steps above to fix any remaining problems.

After that, you may use `mi convert` to switch the data to a more convenient style. We generally start with style wide,

    . mi convert wide

Do not switch to wide, however, if you have any super-varying variables. Try flong instead:

    . mi convert flong

Whichever style you choose, if you get an insufficient-memory error, you will have to either increase the amount of memory dedicated to Stata or use these data in the more inconvenient, but perfectly workable, flongsep style. Concerning increasing memory, see *Converting from flongsep* in [MI] **mi convert**. Concerning the workability of flongsep, see *Advice for using flongsep* in [MI] **styles**.

We said to perform the checks above before using `mi convert`. It is, however, safe to convert the just-imported flongsep data to flong, perform the checks, and then convert to the desired form. The checks will run more quickly if you convert to flong first.

You can vary how you perform the checks. The logic underlying our recommendations is as follows:

- It is possible that you did not specify all the imputed and passive variables when you imported the data, perhaps due to errors in the data's documentation. It is also possible that there are errors in the data that you imported. It is worth checking.

- As long as the imported data are recorded in the flongsep or flong style, unregistered variables will appear exactly as they appeared in the original source. It is only when the data are converted to the wide or mlong style that assumptions about the structure of the data are exploited to save memory. Thus you need to perform checks before converting the data to the more convenient wide or mlong style.

- If you find errors, you could go back and reimport the data correctly, but it is easier to use `mi register` after the fact. When you type `mi register` you are not only informing `mi` about how to deal with the variable but also asking `mi register` to examine the variable and fix any problems given its new registration status.

References

Royston, P. 2004. Multiple imputation of missing values. *Stata Journal* 4: 227–241.

——. 2005a. Multiple imputation of missing values: Update. *Stata Journal* 5: 188–201.

——. 2005b. Multiple imputation of missing values: Update of ice. *Stata Journal* 5: 527–536.

Also see

[MI] **intro** — Introduction to mi

[MI] **mi import nhanes1** — Import NHANES-format data into mi

[MI] **mi import ice** — Import ice-format data into mi

[MI] **mi import flong** — Import flong-like data into mi

[MI] **mi import flongsep** — Import flongsep-like data into mi

[MI] **mi import wide** — Import wide-like data into mi

[MI] **styles** — Dataset styles

Title

mi import flong — Import flong-like data into mi

Syntax

mi import flong, *required_options* [*true_options*]

required_options	description
m(*varname*)	name of variable containing m
id(*varlist*)	identifying variable(s)

true_options	description
imputed(*varlist*)	imputed variables to be registered
passive(*varlist*)	passive variables to be registered
clear	okay to replace unsaved data in memory

Menu

Statistics > Multiple imputation

Description

mi import flong imports flong-like data, that is, data in which $m = 0, m = 1, \ldots, m = M$ are all recorded in one .dta dataset.

mi import flong converts the data to mi flong style. The data are mi set.

Options

m(*varname*) and id(*varlist*) are required. m(*varname*) specifies the variable that takes on values 0, 1, ..., M, the variable that identifies observations corresponding to $m = 0, m = 1, \ldots, m = M$. *varname* = 0 identifies the original data, *varname* = 1 identifies $m = 1$, and so on.

id(*varlist*) specifies the variable or variables that uniquely identify observations within m().

imputed(*varlist*) and passive(*varlist*) are truly optional options, although it would be unusual if imputed() were not specified.

imputed(*varlist*) specifies the names of the imputed variables.

passive(*varlist*) specifies the names of the passive variables, if any.

clear specifies that it is okay to replace the data in memory even if they have changed since they were saved to disk. Remember, mi import flong starts with flong-like data in memory and ends with mi flong data in memory.

Remarks

The procedure to convert flong-like data to mi flong is this:

1. use the unset data.

2. Issue the mi import flong command.

3. Perform the checks outlined in *Using mi import nhanes1, ice, flong, and flongsep* of [MI] **mi import**.

4. Use mi convert (see [MI] **mi convert**) to convert the data to a more convenient style, such as wide or mlong.

For instance, you have the following unset data:

        . use http://www.stata-press.com/data/r11/ourunsetdata
        . list, separator(2)

	m	subject	a	b	c
1.	0	101	1	2	3
2.	0	102	4	.	.
3.	1	101	1	2	3
4.	1	102	4	4.5	8.5
5.	2	101	1	2	3
6.	2	102	4	5.5	9.5

You are told that these data contain the original data (m = 0) and two imputations (m = 1 and m = 2), that variable b is imputed, and that variable c is passive and in fact equal to a + b. These are the same data discussed in [MI] **styles**, but in unset form.

The fact that these data are nicely sorted is irrelevant. To import these data, type

        . mi import flong, m(m) id(subject) imputed(b) passive(c)

These data are short enough that we can list the result:

        . list, separator(2)

	m	subject	a	b	c	_mi_m	_mi_id	_mi_miss
1.	0	101	1	2	3	0	1	0
2.	0	102	4	.	.	0	2	1
3.	1	101	1	2	3	1	1	.
4.	1	102	4	4.5	8.5	1	2	.
5.	2	101	1	2	3	2	1	.
6.	2	102	4	5.5	9.5	2	2	.

We will now perform the checks outlined in *Using mi import nhanes1, ice, flong, and flongsep* of [MI] **mi import**, which are to run mi describe and mi varying to verify that variables are registered correctly:

```
. mi describe
   Style:  flong
           last mi update 15apr2009 11:54:58, 0 seconds ago
   Obs.:   complete       1
           incomplete     1     (M = 2 imputations)
           ─────────────────
           total          2
   Vars.:  imputed: 1; b(1)
           passive: 1; c(1)
           regular: 0
           system:  3; _mi_m _mi_id _mi_miss
           (there are 3 unregistered variables; m subject a)
. mi varying
                Possible problem    variable names
           ─────────────────────────────────────────
                imputed nonvarying:  (none)
                passive nonvarying:  (none)
               unregistered varying: (none)
          *unregistered super/varying: (none)
           unregistered super varying: m

 * super/varying means super varying but would be varying if registered as
   imputed; variables vary only where equal to soft missing in m=0.
```

We discover that unregistered variable m is super varying (see [MI] **Glossary**). Here we no longer need m, so we will drop the variable and rerun `mi varying`. We will find that there are no remaining problems, so we will convert our data to our preferred wide style:

```
. drop m
. mi varying
                Possible problem    variable names
           ─────────────────────────────────────────
                imputed nonvarying:  (none)
                passive nonvarying:  (none)
               unregistered varying: (none)
          *unregistered super/varying: (none)
           unregistered super varying: (none)

 * super/varying means super varying but would be varying if registered as
   imputed; variables vary only where equal to soft missing in m=0.
. mi convert wide, clear
. list
```

	subject	a	b	c	_mi_miss	_1_b	_1_c	_2_b	_2_c
1.	101	1	2	3	0	2	3	2	3
2.	102	4	.	.	1	4.5	8.5	5.5	9.5

Also see

[MI] **intro** — Introduction to mi

[MI] **mi import** — Import data into mi

Title

mi import flongsep — Import flongsep-like data into mi

Syntax

    mi import flongsep *name*, *required_options* [*true_options*]

where *name* is the name of the flongsep data to be created.

required_options	description
using(*filenamelist*)	input filenames for $m=1$, $m=2$, ...
id(*varlist*)	identifying variable(s)

Note: use the input file for $m=0$ before issuing mi import flongsep.

true_options	description
imputed(*varlist*)	imputed variables to be registered
passive(*varlist*)	passive variables to be registered
clear	okay to replace unsaved data in memory

Menu

Statistics > Multiple imputation

Description

    mi import flongsep imports flongsep-like data, that is, data in which $m=0$, $m=1$, ..., $m=M$ are each recorded in separate .dta datasets.

    mi import flongsep converts the data to mi flongsep and mi sets the data.

Options

    using(*filenamelist*) is required; it specifies the names of the .dta datasets containing $m=1$, $m=2$, ..., $m=M$. The dataset corresponding to $m=0$ is not specified; it is to be in memory at the time the mi import flongsep command is given.

    The filenames might be specified as

        using(ds1 ds2 ds3 ds4 ds5)

    which states that $m=1$ is in file ds1.dta, $m=2$ is in file ds2.dta, ..., and $m=5$ is in file ds5.dta. Also, {#-#} is understood, so the above could just as well be specified as

        using(ds{1-5})

The braced numeric range may appear anywhere in the name, and thus

   using(ds{1-5}imp)

would mean that ds1imp.dta, ds2imp.dta, ..., ds5imp.dta contain $m = 1$, $m = 2$, ..., $m = 5$.

Alternatively, a comma-separated list can appear inside the braces. Filenames dsfirstm.dta, dssecondm.dta, ..., dsfifthm.dta can be specified as

   using(ds{first,second,third,fourth,fifth}m)

Filenames can be specified with or without the .dta suffix and may be enclosed in quotes if they contain special characters.

id(*varlist*) is required; it specifies the variable or variables that uniquely identify the observations in each dataset. The coding must be the same across datasets.

imputed(*varlist*) and passive(*varlist*) are truly optional options, although it would be unusual if imputed() were not specified.

imputed(*varlist*) specifies the names of the imputed variables.

passive(*varlist*) specifies the names of the passive variables.

clear specifies that it is okay to replace the data in memory even if they have changed since they were saved to disk.

Remarks

The procedure to convert flongsep-like data to mi flongsep is this:

1. use the dataset corresponding to $m = 0$.

2. Issue the mi import flongsep *name* command, where *name* is the name of the mi flongsep data to be created.

3. Perform the checks outlined in *Using mi import nhanes1, ice, flong, and flongsep* of [MI] **mi import**.

4. Use mi convert (see [MI] **mi convert**) to convert the data to a more convenient style such as wide, mlong, or flong.

For instance, you have been given the unset datasets imorig.dta, im1.dta, and im2.dta. You are told that these datasets contain the original data and two imputations, that variable b is imputed, and that variable c is passive and in fact equal to a + b. Here are the datasets:

```
. use http://www.stata-press.com/data/r11/imorig
. list
```

	subject	a	b	c
1.	101	1	2	3
2.	102	4	.	.

```
. use http://www.stata-press.com/data/r11/im1
. list
```

	subject	a	b	c
1.	101	1	2	3
2.	102	4	4.5	8.5

```
. use http://www.stata-press.com/data/r11/im2
. list
```

	subject	a	b	c
1.	101	1	2	3
2.	102	4	5.5	9.5

These are the same data discussed in [MI] **styles**, but unset.

The fact that these datasets are nicely sorted is irrelevant. To import these datasets, you type

```
. use http://www.stata-press.com/data/r11/imorig
. mi import flongsep mymi, using(im1 im2) id(subject) imputed(b) passive(c)
```

We will now perform the checks outlined in *Using mi import nhanes1, ice, flong, and flongsep* of [MI] **mi import**, which are to run mi describe and mi varying to verify that variables are registered correctly:

```
. mi describe
  Style:  flongsep mymi
          last mi update 15apr2009 13:21:08, 0 seconds ago
  Obs.:   complete         1
          incomplete       1   (M = 2 imputations)
          ─────────────
          total            2
  Vars.:  imputed:  1; b(1)
          passive:  1; c(1)
          regular:  0
          system:   2; _mi_id _mi_miss
          (there are 2 unregistered variables; subject a)
. mi varying

               Possible problem    variable names
          ──────────────────────────────────────────
                imputed nonvarying: (none)
                passive nonvarying: (none)
             unregistered varying: (none)
       *unregistered super/varying: (none)
        unregistered super varying: (none)
          ──────────────────────────────────────────
       * super/varying means super varying but would be varying if registered as
         imputed; variables vary only where equal to soft missing in m=0.
```

mi varying reported no problems. We finally convert to our preferred wide style:

. mi convert wide, clear

. list

	subject	a	b	c	_mi_miss	_1_b	_1_c	_2_b	_2_c
1.	101	1	2	3	0	2	3	2	3
2.	102	4	.	.	1	4.5	8.5	5.5	9.5

We are done with the converted data in flongsep format, so we will erase the files:

```
. mi erase mymi
(files mymi.dta _1_mymi.dta _2_mymi.dta erased)
```

Also see

[MI] **intro** — Introduction to mi

[MI] **mi import** — Import data into mi

Title

> **mi import ice** — Import ice-format data into mi

Syntax

    mi import ice [, options]

options	description
automatic	register variables automatically
imputed(*varlist*)	imputed variables to be registered
passive(*varlist*)	passive variables to be registered
clear	okay to replace unsaved data

Menu

Statistics > Multiple imputation

Description

mi import ice converts the data in memory to mi data, assuming the data in memory are in ice format. See Royston (2004, 2005a, 2005b) for a description of ice.

mi import ice converts the data to mi style flong. The data are mi set.

Options

automatic determines the identity of the imputed variables automatically. Use of this option is recommended.

imputed(*varlist*) specifies the names of the imputed variables. This option may be used with automatic, in which case automatic is taken to mean automatically determine the identity of imputed variables in addition to the imputed() variables specified. It is difficult to imagine why one would want to do this.

passive(*varlist*) specifies the names of the passive variables. This option may be used with automatic and usefully so. automatic cannot distinguish imputed variables from passive variables, so it assumes all variables that vary are imputed. passive() allows you to specify the subset of varying variables that are passive.

Concerning the above options, if none are specified, all variables are left unregistered in the result. You can then use mi varying to determine the varying variables and use mi register to register them appropriately; see [MI] **mi varying** and [MI] **mi set**. If you follow this approach, remember to register imputed variables before registering passive variables.

clear specifies that it is okay to replace the data in memory even if they have changed since they were last saved to disk. Remember, mi import ice starts with ice data in memory and ends with mi data in memory.

Remarks

The procedure to convert `ice` data to `mi` flong is

1. use the `ice` data.

2. Issue the `mi import ice` command, preferably with the `automatic` option and perhaps with the `passive()` option, too, although it really does not matter if passive variables are registered as imputed, so long as they are registered.

3. Perform the checks outlined in *Using mi import nhanes1, ice, flong, and flongsep* of [MI] **mi import**.

4. Use `mi convert` (see [MI] **mi convert**) to convert the data to a more convenient style such as wide or mlong.

For instance, you have the following ice data:

```
. use http://www.stata-press.com/data/r11/icedata
. list, separator(2)
```

	_mj	_mi	a	b	c
1.	0	1	1	2	3
2.	0	2	4	.	.
3.	1	1	1	2	3
4.	1	2	4	4.5	8.5
5.	2	1	1	2	3
6.	2	2	4	5.5	9.5

_mj and _mi are `ice` system variables. These data contain the original data and two imputations. Variable b is imputed, and variable c is passive and in fact equal to a + b. These are the same data discussed in [MI] **styles**, but in `ice` format.

The fact that these data are nicely sorted is irrelevant. To import these data, you type

```
. mi import ice, automatic
(1 m=0 obs. now marked as incomplete)
```

although it would be even better if you typed

```
. mi import ice, automatic passive(c)
(1 m=0 obs. now marked as incomplete)
```

With the first command, both b and c will be registered as imputed. With the second, c will instead be registered as passive. Whether c is registered as imputed or passive makes no difference statistically.

These data are short enough that we can list the result:

. list, separator(2)

	a	b	c	_mi_m	_mi_id	_mi_miss
1.	1	2	3	0	1	0
2.	4	.	.	0	2	1
3.	1	2	3	1	1	.
4.	4	4.5	8.5	1	2	.
5.	1	2	3	2	1	.
6.	4	5.5	9.5	2	2	.

We will now perform the checks outlined in *Using mi import nhanes1, ice, flong, and flongsep* of [MI] **mi import**, which are to run mi describe and mi varying to verify that variables are registered correctly:

```
. mi describe
  Style:    flong
            last mi update 15apr2009 11:54:58, 0 seconds ago
  Obs.:     complete         1
            incomplete       1   (M = 2 imputations)
            ─────────────────────
            total            2
  Vars.:    imputed: 1; b(1)
            passive: 1; c(1)
            regular: 0
            system:  3; _mi_m _mi_id _mi_miss
            (there is one unregistered variable; a)
. mi varying

                  Possible problem    variable names
        ──────────────────────────────────────────────
              imputed nonvarying:     (none)
              passive nonvarying:     (none)
              unregistered varying:   (none)
        *unregistered super/varying:  (none)
          unregistered super varying: (none)
        ──────────────────────────────────────────────

    * super/varying means super varying but would be varying if registered as
      imputed; variables vary only where equal to soft missing in m=0.
```

We find that there are no remaining problems, so we convert our data to our preferred wide style:

. mi convert wide, clear

. list

	a	b	c	_mi_miss	_1_b	_1_c	_2_b	_2_c
1.	1	2	3	0	2	3	2	3
2.	4	.	.	1	4.5	8.5	5.5	9.5

References

Royston, P. 2004. Multiple imputation of missing values. *Stata Journal* 4: 227–241.

——. 2005a. Multiple imputation of missing values: Update. *Stata Journal* 5: 188–201.

——. 2005b. Multiple imputation of missing values: Update of ice. *Stata Journal* 5: 527–536.

Also see

[MI] **intro** — Introduction to mi

[MI] **mi import** — Import data into mi

Title

mi import nhanes1 — Import NHANES-format data into mi

Syntax

mi import nhanes1 *name*, *required_options* [*true_options odd_options*]

where *name* is the name of the flongsep data to be created.

required_options	description
using(*filenamelist*)	input filenames for $m = 1, m = 2, \ldots$
id(*varlist*)	identifying variable(s)

Note: use the input file for $m=0$ before issuing mi import nhanes1.

true_options	description
uppercase	prefix and suffix in uppercase
clear	okay to replace unsaved data in memory

odd_options	description
nacode(*#*)	not applicable code; default is 0
obscode(*#*)	observed code; default is 1
impcode(*#*)	imputed code; default is 2
impprefix("*string*" "*string*")	variable prefix; default is "" ""
impsuffix("*string*" "*string*")	variable suffix; default is "if" "mi"

Note: The *odd_options* are not specified unless you need to import data that are nhanes1-like but not really nhanes1 format.

Menu

Statistics > Multiple imputation

Description

mi import nhanes1 imports data recorded in the format used by the National Health and Nutrition Examination Survey (NHANES) produced by the National Center for Health Statistics (NCHS) of the U.S. Centers for Disease Control and Prevention (CDC); see http://www.cdc.gov/nchs/nhanes/nh3data.htm.

Options

using(*filenamelist*) is required; it specifies the names of the .dta datasets containing $m = 1$, $m = 2, \ldots, m = M$. The dataset corresponding to $m = 0$ is not specified; it is to be in memory at the time the mi import nhanes1 command is given.

The filenames might be specified as

> using(nh1 nh2 nh3 nh4 nh5)

which states that $m=1$ is in file nh1.dta, $m=2$ is in file nh2.dta, ..., and $m=5$ is in file nh5.dta. Also, {#-#} is understood, so the files could just as well be specified as

> using(nh{1-5})

The braced numeric range may appear anywhere in the name, and thus

> using(nh{1-5}imp)

would mean that nh1imp.dta, nh2imp.dta, ..., nh5imp.dta contain $m=1$, $m=2$, ..., $m=5$.

Alternatively, a comma-separated list can appear inside the braces. Filenames nhfirstm.dta, nhsecondm.dta, ..., nhfifthm.dta can be specified as

> using(nh{first,second,third,fourth,fifth}m)

Filenames can be specified with or without the .dta suffix and must be enclosed in quotes if they contain special characters.

id(*varlist*) is required and is usually specified as id(seqn) or id(SEQN) depending on whether your variable names are in lowercase or uppercase. id() specifies the variable or variables that uniquely identify the observations in each dataset. Per the nhanes1 standard, the variable should be named seqn or SEQN.

uppercase is optional; it specifies that the variable suffixes IF and MI of the nhanes1 standard are in uppercase. The default is lowercase. (More correctly, when generalizing beyond nhanes1 format, the uppercase option specifies that all prefixes and suffixes are in uppercase.)

nacode(*#*), obscode(*#*), and impcode(*#*) are optional and are never specified when reading true nhanes1 data. The defaults nacode(0), obscode(1), and impcode(2) correspond to the nhanes1 definition. These options allow changing the codes for not applicable, observed, and imputed.

impprefix("*string*" "*string*") and impsuffix("*string*" "*string*") are optional and are never specified when reading true nhanes1 data. The defaults impprefix("" "") and impsuffix("if" "mi") correspond to the nhanes1 definition. These options allow setting different prefixes and suffixes.

clear specifies that it is okay to replace the data in memory even if they have changed since they were saved to disk. Remember, mi import nhanes1 starts with the first of the NHANES data in memory and ends with mi data in memory.

Remarks

Remarks are presented under the following headings:

> *Description of the nhanes1 format*
> *Importing nhanes1 data*

Description of the nhanes1 format

Nhanes1 is not really an official format; it is the format used for a particular dataset distributed by NCHS. Because there currently are no official or even informal standards for multiple-imputation data, perhaps the method used by the NCHS for NHANES will catch on, so we named it nhanes1. We included the 1 on the end of the name in case the format is modified.

Data in nhanes1 format consist of a collection of $M+1$ separate files. The first file contains the original data. The remaining M files contain the imputed values for $m=1, m=2, \ldots, m=M$.

The first file contains a variable named seqn containing a sequence number. The file also contains other variables that comprise the nonimputed variables. Imputed variables, however, have their names suffixed with IF, standing for imputation flag, and those variables contain 1s, 2s, and 0s. 1 means that the value of the variable in that observation was observed, 2 means that the value was missing, and 0 means not applicable. Think of 0 as being equivalent to hard missing. The value is not observed for good reason and therefore was not imputed.

The remaining M files contain seqn and the imputed variables themselves. In these files, unobserved values are imputed. This time, imputed variable names are suffixed with MI.

Here is an example:

```
. use http://www.stata-press.com/data/r11/nhorig
. list
```

	seqn	a	bIF	cIF
1.	1	11	1	1
2.	2	14	2	2

The above is the first of the $M+1$ datasets. The seqn variable is the sequence number. The a variable is a regular variable; we know that because the name does not end in IF. The b and c variables are imputed, and this dataset contains their imputation flags. Both variables are observed in the first observation and unobserved in the second.

Here is the corresponding dataset for $m=1$:

```
. use http://www.stata-press.com/data/r11/nh1
. list
```

	seqn	bMI	cMI
1.	1	2	3
2.	2	4.5	8.5

This dataset states that in $m=1$, b is equal to 2 and 4.5 and c is equal to 3 and 8.5.

We are about to show you the dataset for $m=2$. Even before looking at it, however, we know that 1) it will have two observations; 2) it will have the seqn variable containing 1 and 2; 3) it will have two more variables named bMI and cMI; and 4) bMI will be equal to 2 and cMI will be equal to 3 in observations corresponding to seqn = 1. We know the last because in the first dataset, we learned that b and c were observed in seqn = 1.

```
. use http://www.stata-press.com/data/r11/nh2
. list
```

	seqn	a	bMI	cMI
1.	1	11	2	3
2.	2	14	5.5	9.5

Importing nhanes1 data

The procedure to import nhanes1 data is this:

1. use the dataset corresponding to $m = 0$; see [D] **use**.
2. Issue `mi import nhanes1` *name* ..., where *name* is the name of the mi flongsep dataset to be created.
3. Perform the checks outlined in *Using mi import nhanes1, ice, flong, and flongsep* of [MI] **mi import**.
4. Use `mi convert` (see [MI] **mi convert**) to convert the data to a more convenient style such as wide, mlong, or flong.

To import the nhorig.dta, nh1.dta, and nh2.dta datasets described in the section above, we will specify `mi import nhanes1`'s `uppercase` option because the suffixes were in uppercase. We type

```
. use http://www.stata-press.com/data/r11/nhorig
. mi import nhanes1 mymi, using(nh1 nh2) id(seqn) uppercase
```

The lack of any error message means that we have successfully converted nhanes1-format files nhorig.dta, nh1.dta, and nh2.dta to mi flongsep files mymi.dta, _1_mymi.dta, and _2_mymi.dta.

We will now perform the checks outlined in *Using mi import nhanes1, ice, flong, and flongsep* of [MI] **mi import**, which are to run `mi describe` and `mi varying` to verify that variables are registered correctly:

```
. mi describe
  Style:  flongsep mymi
          last mi update 16apr2009 12:58:46, 0 seconds ago
   Obs.:  complete           1
          incomplete         1   (M = 2 imputations)
          ─────────────────────
          total              2
   Vars.: imputed:  2; b(1) c(1)
          passive:  0
          regular:  0
          system:   2; _mi_id _mi_miss
          (there are 2 unregistered variables; seqn a)
```

```
. mi varying
                Possible problem   variable names
             imputed nonvarying:   (none)
             passive nonvarying:   (none)
          unregistered varying:    (none)
       *unregistered super/varying: (none)
         unregistered super varying: (none)

    * super/varying means super varying but would be varying if registered as
      imputed; variables vary only where equal to soft missing in m=0.
```

mi varying reported no problems.

We finally convert to style flong, although in real life we would choose styles mlong or wide. We are choosing flong because it is more readable:

```
. mi convert flong, clear

. list, separator(2)
```

	seqn	a	b	c	_mi_id	_mi_miss	_mi_m
1.	1	11	2	3	1	0	0
2.	2	14	.	.	2	1	0
3.	1	11	2	3	1	.	1
4.	2	14	4.5	8.5	2	.	1
5.	1	11	2	3	1	.	2
6.	2	14	5.5	9.5	2	.	2

The flong data are in memory. We are done with the converted data in flongsep format, so we erase the files:

```
. mi erase mymi
(files mymi.dta _1_mymi.dta _2_mymi.dta erased)
```

Also see

[MI] **intro** — Introduction to mi

[MI] **mi import** — Import data into mi

Title

mi import wide — Import wide-like data into mi

Syntax

    mi import wide [, *options*]

options	description
<u>imp</u>uted(*mvlist*)	imputed variables
<u>pas</u>sive(*mvlist*)	passive variables
<u>dups</u>ok	allow variable to be posted repeatedly
drop	drop imputed and passive after posting
clear	okay to replace unsaved data in memory

See description of options below for definition of *mvlist*.

Menu

Statistics > Multiple imputation

Description

    mi import wide imports wide-like data, that is, data in which $m = 0$, $m = 1$, ..., $m = M$ values of imputed and passive variables are recorded in separate variables.

    mi import wide converts the data to mi wide style and mi sets the data.

Options

    imputed(*mvlist*) and passive(*mvlist*) specify the imputed and passive variables.

    For instance, if the data had two imputed variables, x and y; x and y contained the $m = 0$ values; the corresponding $m = 1$, $m = 2$, and $m = 3$ values of x were in x1, x2, and x3; and the corresponding values of y were in y1, y2, and y3, then the imputed() option would be specified as

        imputed(x=x1 x2 x3 y=y1 y2 y3)

    If variable y2 were missing from the data, you would specify

        imputed(x=x1 x2 x3 y=y1 . y3)

    The same number of imputations must be specified for each variable.

    dupsok specifies that it is okay if you specify the same variable name for two different imputations. This would be an odd thing to do, but if you specify dupsok, then you can specify

        imputed(x=x1 x1 x3 y=y1 y2 y3)

    Without the dupsok option, the above would be treated as an error.

drop specifies that the original variables containing values for $m = 1, m = 2, \ldots, m = M$ are to be dropped from the data once `mi import wide` has recorded the values. This option is recommended.

clear specifies that it is okay to replace the data in memory even if they have changed since they were last saved to disk.

Remarks

The procedure to convert wide-like data to `mi` wide style is this:

1. `use` the unset data.
2. Issue the `mi import wide` command.
3. Use `mi describe` (see [MI] **mi describe**) and `mi varying` (see [MI] **mi varying**) to verify that the result is as you anticipated.
4. Optionally, use `mi convert` (see [MI] **mi convert**) to convert the data to what you consider a more convenient style.

For instance, you have been given unset dataset `wi.dta` and have been told that it contains variables a, b, and c; that variable b is imputed and contains $m = 0$ values; that variables b1 and b2 contain the $m = 1$ and $m = 2$ values; that variable c is passive (equal to a + b) and contains $m = 0$ values; and that variables c1 and c2 contain the corresponding $m = 1$ and $m = 2$ values. Here are the data:

```
. use http://www.stata-press.com/data/r11/wi
(mi prototype)

. list
```

	a	b	c	b1	b2	c1	c2
1.	1	2	3	2	2	3	3
2.	4	.	.	4.5	5.5	8.5	9.5

These are the same data discussed in [MI] **styles**. To import these data, type

```
. mi import wide, imputed(b=b1 b2  c=c1 c2) drop
```

These data are short enough that we can list the result:

```
. list
```

	a	b	c	_mi_miss	_1_b	_2_b	_1_c	_2_c
1.	1	2	3	0	2	2	3	3
2.	4	.	.	1	4.5	5.5	8.5	9.5

(*Continued on next page*)

Returning to the procedure, we run `mi describe` and `mi varying` on the result:

```
. mi describe
  Style: wide
         last mi update 16apr2009 08:51:09, 0 seconds ago
  Obs.:  complete                1
         incomplete              1  (M = 2 imputations)
         ─────────────────────────
         total                   2
  Vars.: imputed:  2; b(1) c(1)
         passive:  0
         regular:  0
         system:   1; _mi_miss
         (there is one unregistered variable; a)
. mi varying

                    Possible problem    variable names
         ─────────────────────────────────────────────
                   imputed nonvarying:   (none)
                   passive nonvarying:   (none)
```

Perhaps you would prefer seeing these data in flong style:

```
. mi convert flong, clear
. list, separator(2)

     ┌──────────────────────────────────────────┐
     │ a     b     c    _mi_miss  _mi_m  _mi_id │
     ├──────────────────────────────────────────┤
  1. │ 1     2     3       0        0       1   │
  2. │ 4     .     .       1        0       2   │
     ├──────────────────────────────────────────┤
  3. │ 1     2     3       .        1       1   │
  4. │ 4    4.5   8.5      .        1       2   │
     ├──────────────────────────────────────────┤
  5. │ 1     2     3       .        2       1   │
  6. │ 4    5.5   9.5      .        2       2   │
     └──────────────────────────────────────────┘
```

Also see

[MI] **intro** — Introduction to mi

[MI] **mi import** — Import data into mi

Title

> **mi impute** — Impute missing values

Syntax

> mi impute *method* ... [, *impute_options* ...]

method	description
Univariate	
regress	linear regression for a continuous variable
pmm	predictive mean matching for a continuous variable
logit	logistic regression for a binary variable
ologit	ordinal logistic regression for an ordinal variable
mlogit	multinomial logistic regression for a nominal variable
Multivariate	
monotone	sequential imputation using a monotone-missing pattern
mvn	multivariate normal regression

impute_options	description
Main	
* add(#)	specify number of imputations to add; required when no imputations exist
* replace	replace imputed values in existing imputations
rseed(#)	specify random-number seed
double	save imputed values in double precision; the default is to save them as float
Reporting	
dots	display dots as imputations are performed
noisily	display intermediate output
Advanced	
force	proceed with imputation even when missing imputed values are encountered
† noupdate	do not perform mi update; see [MI] **noupdate option**

*add(#) is required when no imputations exist; add(#) or replace is required if imputations exist.

†noupdate does not appear in the dialog box.

You must mi set your data before using mi impute; see [MI] **mi set**.

mi impute — Impute missing values

Menu

Statistics > Multiple imputation

Description

mi impute fills in missing values (.) of a single variable or of multiple variables using the specified method. The available methods (by variable type and missing-data pattern) are summarized in the tables below.

Single imputation variable (univariate imputation)

Pattern	Type	Imputation method
	continuous	regress, pmm
always monotone	binary	logit
	categorical	ologit, logit

Multiple imputation variables (multivariate imputation)

Pattern	Type	Imputation method
monotone missing	mixture	monotone
arbitrary missing	continuous	mvn

The suggested reading order of mi impute's subentries is

[MI] **mi impute regress**

[MI] **mi impute pmm**

[MI] **mi impute logit**

[MI] **mi impute mlogit**

[MI] **mi impute ologit**

[MI] **mi impute monotone**

[MI] **mi impute mvn**

Options

Main

add(#) specifies the number of imputations to add to the mi data. This option is required if there are no imputations in the data. If imputations exist, then add() is optional. The total number of imputations cannot exceed 1,000.

replace specifies to replace existing imputed values with new ones. One of replace or add() must be specified when mi data already have imputations ($M > 0$).

rseed(#) sets the random-number seed. This option can be used to reproduce results. rseed(#) is equivalent to typing set seed # prior to calling mi impute; see [R] **set seed**.

double specifies that the imputed values be stored as doubles. By default, they are stored as floats. mi impute stores the values using double or float precision only when necessary. For example, if the logit method is used, the imputed values are stored as bytes.

Reporting

dots specifies to display dots as imputations are successfully completed. An **x** is displayed if any of the specified imputation variables still have missing values.

noisily specifies that intermediate output from mi impute be displayed.

Advanced

force specifies to proceed with imputation even when missing imputed values are encountered. By default, mi impute terminates with error if missing imputed values are encountered.

noupdate in some cases suppresses the automatic mi update this command might perform; see [MI] **noupdate option**. This option is rarely used.

Remarks

Remarks are presented under the following headings:

> *Imputation methods*
> *Imputation modeling*
> > *Model building*
> > *Outcome variables*
> > *Transformations*
> > *Categorical variables*
> > *Convergence of MCMC*
>
> *Using mi impute*
> *Univariate imputation*
> *Multivariate imputation*
> *Imputing on subsamples*
> *Imputation and estimation samples*
> *Imputing transformations of imputed variables*

Imputation methods

mi impute supports both univariate and multivariate imputation.

Univariate imputation is used to impute a single variable. It can be used repeatedly to impute multiple variables, and this can be justified when the variables are independent and will be used in separate analyses. To impute a single variable, you can choose from the following methods: regress, pmm, logit, ologit, and mlogit.

For a continuous variable, either regress or pmm can be used (for example, Rubin [1987] and Schenker and Taylor [1996]). For a binary variable, logit can be used (Rubin 1987). For a categorical variable, ologit can be used to impute missing categories if they are ordered, and mlogit can be used to impute missing categories if they are unordered (Raghunathan et al. 2001). Also see van Buuren (2007) for a detailed list of univariate imputation methods.

In practice, multiple variables usually must be imputed simultaneously, and that requires using a multivariate imputation method. The choice of an imputation method in this case depends on the pattern of missing values.

If variables follow a monotone-missing pattern (see *Patterns of missing data* in [MI] **intro substantive**), they can be imputed sequentially using independent univariate conditional distributions, which is implemented in the `monotone` method. A separate univariate imputation model can be specified for each imputation variable, which allows simultaneous imputation of variables of different types (Rubin 1987).

When a pattern of missing values is arbitrary, an iterative method is required to fill in missing values. The `mvn` method uses multivariate normal data augmentation to impute missing values of continuous imputation variables (Schafer 1997). Allison (2001), for example, also uses this method to impute binary and categorical variables.

Imputation modeling

As discussed in [MI] **intro substantive**, imputation modeling is important to obtain proper imputations. Imputation modeling is not confined to the specification of an imputation method and an imputation model. It also requires careful consideration of how to handle complex data structures, such as survey or longitudinal data, and how to preserve existing relationships in the data during the imputation step. Rubin (1987), Schafer (1997), Allison (2001), Graham (2009), Royston (2007), and others provide guidelines about imputation modeling. We summarize some of them below.

Model building

Perhaps the most important component of imputation modeling is the construction of an imputation model that preserves all the main characteristics of the observed data. This includes the following:

1. Use as many predictors as possible in the model to avoid making incorrect assumptions about the relationship between the variables. Omitting key predictors from the imputation model may lead to biased estimates for these predictors in the analysis. On the other hand, including insignificant predictors will result in less efficient yet still statistically valid results.

2. Include design variables representing the structure of the data in your imputation model. For example, sampling weights, strata and cluster identifiers of survey data, repeated-measures identifiers of longitudinal data must be included in the imputation model.

3. Include interactions of variables (or impute missing values separately using different subsamples; see *Imputing on subsamples*) to preserve higher-order dependencies.

The imputation model must be compatible with any model that can be used for the analysis. If variable X is to be included in the analysis model, it should also be used in the imputation model. If the analysis model estimates a correlation of X_1 and X_2, then both variables should be present in the imputation model. Accordingly, the outcome variable should always be present in the imputation model. Also, in addition to all the variables that may be used in the analysis model, you should include any auxiliary variables that may contain information about missing data. This will improve the quality of the imputed values.

Outcome variables

Imputing outcome variables received special attention in the literature because of the controversy about whether they should be imputed. As we already mentioned, it is important to include the outcome variable in the imputation model to obtain valid results. But what if the outcome variable itself has missing values? Should it be imputed? Should missing values be discarded from the analysis? There is no definitive answer to this question. The answer ultimately comes down to whether the specified imputation model describes the missing data adequately. When the percentage of missing values is low, using an incorrect imputation model may have very little effect on the resulting repeated-imputation inference. With a large fraction of missing observations, a misspecified imputation model may distort the observed relationship between the outcome and predictor variables. In general, with large fractions of missing observations on any variable, the imputed values have more influence on the results, and thus more careful consideration of the imputation probability model is needed.

Transformations

Although the choice of an imputation method may not have significant impact on the results with low fractions of missing data, it may with larger fractions. A number of different imputation methods is available to model various types of imputation variables, from continuous to categorical. However, they in no way cover all possible distributions that imputation variables may have in practice. Often, the imputation variables can be transformed to the scale appropriate for an imputation method. For example, a log transformation can be used for highly skewed continuous variables to make them suitable for imputation using the linear regression method. If desired, the imputed values can be transformed back after the imputation. Transformations are useful when a variable has a restricted range. For instance, a preimputation logit transformation and a postimputation inverse logit transformation can be used to ensure that the imputed values are between 0 and 1.

Categorical variables

To impute one categorical variable, you can use one of the categorical imputation methods: logistic, ordered logistic, or multinomial logistic regressions ([MI] **mi impute logit**, [MI] **mi impute ologit**, [MI] **mi impute mlogit**). These methods can also be used to impute multiple categorical variables within the monotone-missing data framework ([MI] **mi impute monotone**). With an arbitrary missing-data pattern, the choice for multivariate imputation of categorical variables is limited. For multiple categorical variables with only two categories (binary or dummy variables), a multivariate normal approach ([MI] **mi impute mvn**) can be used to impute missing values and then round the imputed values to 0 if the value is smaller than 0.5, or 1 otherwise. For categorical variables with more than two categories, Allison (2001) describes how to use the normal model to impute missing values.

Convergence of MCMC

With arbitrary missing data, an iterative (MCMC) imputation method is used to simulate from the posterior predictive distribution of the missing data given the observed data ([MI] **mi impute mvn**). In this case, the resulting sequences of simulated parameters (chains) should be examined to verify the convergence of the algorithm. The modeling task may be influenced by the convergence process of the algorithm given the data. For example, a different prior distribution for the model parameters may be needed when some aspects of the model cannot be estimated because of the sparseness of the missing data. See [MI] **mi impute mvn** for how to monitor convergence of MCMC.

Using mi impute

To use `mi impute`, you first `mi set` your data; see [MI] **mi set**. Next you register all variables whose missing values are to be imputed; see [MI] **mi set**.

`mi impute` has two main options: `add()` and `replace`. If you do not have imputations, use `add()` to create them. If you already have imputations, you have three choices:

1. Add new imputations to the existing ones—specify the `add()` option.

2. Add new imputations and also replace the existing ones—specify both the add() and the replace options.

3. Replace existing imputed values—specify the replace option.

add() is required if no imputations exist in the mi data, and either add() or replace must be specified if imputations exist. See *Univariate imputation* for examples. Note that with replace, only imputed values of the specified imputation variables within the specified subsample will be updated.

For reproducibility, use the rseed() option to set the random-number seed, or equivalently, set the seed by using set seed immediately before calling mi impute. If you forget and still have mi impute's saved results in memory, you can retrieve the seed from r(rseed); see *Saved results* below.

By default, mi impute saves the imputed values using float precision. If you need more accuracy, you can specify the double option. Depending on the mi data style, the type of the imputed variable may change in the original data, $m = 0$. For example, if your data are in the mlong (or flong) style and you are imputing a binary variable using the regression method, the type of the variable will become float. If you are using the logistic method, the type of the variable may become byte even if originally your variable was declared as float or int. mi impute will never demote a variable if that would result in loss of precision.

mi impute terminates with error if the imputation procedure results in missing imputed values. This may happen if you include variables containing missing values as predictors in your imputation model. If desired, you can override this behavior with the force option.

mi impute may change the sort order of the data.

Univariate imputation

Consider the heart attack data in which bmi contains missing values, as described in *A brief introduction to MI using Stata* of [MI] **intro substantive**. Here we use the already mi set version of the data with a subset of covariates of interest:

```
. use http://www.stata-press.com/data/r11/mheart1s0
(Fictional heart attack data; bmi missing)
. mi describe
  Style:  mlong
          last mi update 19jun2009 10:50:18, 2 days ago
  Obs.:   complete         132
          incomplete        22   (M = 0 imputations)
          total            154
  Vars.:  imputed: 1; bmi(22)
          passive: 0
          regular: 5; attack smokes age female hsgrad
          system:  3; _mi_m _mi_id _mi_miss
          (there are no unregistered variables)
```

According to mi describe, the mi data style is mlong, and the dataset contains no imputations and 22 incomplete observations. The only registered imputed variable is bmi containing the 22 missing values. The other variables are registered as regular. See [MI] **mi describe** for details.

In the example in [MI] **intro substantive**, we used mi impute regress to impute missing values of bmi. Let's concentrate on the imputation step in more detail here:

```
. mi impute regress bmi attack smokes age female hsgrad, add(20)
Univariate imputation                Imputations  =       20
Linear regression                          added  =       20
Imputed: m=1 through m=20                updated  =        0
                         |      Observations per m
                Variable | complete   incomplete    imputed  |    total
                ---------|------------------------------------|---------
                     bmi |      132           22         22  |      154
```

(complete + incomplete = total; imputed is the minimum across m of the number of filled in observations.)

The above output is common to all imputation methods of mi impute. In the left column, mi impute reports information about which imputation method was used and which imputations were created or updated. The right column contains the total number of imputations, and how many of them are new and how many are updated. The table contains the number of complete, incomplete, and imputed observations, and the total number of observations in the imputation sample, per imputation for each variable (see *Imputation and estimation samples* below). As indicated by the note, complete and incomplete observations sum to the total number of observations. The imputed column reports how many incomplete observations were actually imputed. This number represents the minimum across all imputations used ($m = 1$ through $m = 20$ in our example).

In the above example, we used add(20) to create 20 new imputations. Suppose that we decided that 20 is not enough and we want to add 30 more:

```
. mi impute regress bmi attack smokes age female hsgrad, add(30)
Univariate imputation                Imputations  =       50
Linear regression                          added  =       30
Imputed: m=21 through m=50               updated  =        0
                         |      Observations per m
                Variable | complete   incomplete    imputed  |    total
                ---------|------------------------------------|---------
                     bmi |      132           22         22  |      154
```

(complete + incomplete = total; imputed is the minimum across m of the number of filled in observations.)

The table output is unchanged, but the header reports that total number of imputations is now 50. Thirty new imputations (from $m = 21$ to $m = 50$) were added, and the existing 20 imputations were left unchanged.

Suppose that we decide we want to impute bmi using the predictive mean matching imputation method instead of the regression method. We use mi impute pmm and specify the replace option to update all existing imputations with new ones:

```
. mi impute pmm bmi attack smokes age female hsgrad, replace
Univariate imputation                Imputations  =       50
Predictive mean matching                   added  =        0
Imputed: m=1 through m=50                updated  =       50
                         |      Observations per m
                Variable | complete   incomplete    imputed  |    total
                ---------|------------------------------------|---------
                     bmi |      132           22         22  |      154
```

(complete + incomplete = total; imputed is the minimum across m of the number of filled in observations.)

The header reports that all the 50 existing imputations, from $m = 1$ to $m = 50$, are replaced with new ones.

Later we decide to use more nearest neighbors than the default with `mi impute pmm` and also add 15 more imputations. We can do the latter by combining `replace` and `add()`. We specify `replace` to update the existing imputations with imputations from PMM with 3 nearest neighbors (`knn(3)`) and use `add(15)` to add 15 more imputations.

```
. mi impute pmm bmi attack smokes age female hsgrad, add(15) replace knn(3) dots
Imputing m=1 through m=65:
.........10.........20.........30.........40.........50.........60..... done

Univariate imputation              Imputations =       65
Predictive mean matching                 added =       15
Imputed: m=1 through m=65              updated =       50
```

	Observations per m			
Variable	complete	incomplete	imputed	total
bmi	132	22	22	154

```
(complete + incomplete = total; imputed is the minimum across m
 of the number of filled in observations.)
```

The header reports a total of 65 imputations, among which 15 are new and 50 are updated. In this example, we also used the `dots` option to see the imputation progress. This option is useful with larger datasets to monitor the imputation progress.

See *Imputing on subsamples* for other usage of `add()` and `replace`.

Multivariate imputation

Consider the heart attack data in which both `bmi` and `age` contain missing values. Again we will use data that have already been `mi set`.

```
. use http://www.stata-press.com/data/r11/mheart5s0
(Fictional heart attack data; bmi and age missing)
. mi describe
  Style:   mlong
           last mi update 19jun2009 10:50:18, 2 days ago
  Obs.:    complete          126
           incomplete         28   (M = 0 imputations)

           total             154
  Vars.:   imputed:  2; bmi(28) age(12)
           passive:  0
           regular:  4; attack smokes female hsgrad
           system:   3; _mi_m _mi_id _mi_miss
           (there are no unregistered variables)
```

There are 28 incomplete observations in the dataset. The `bmi` variable contains 28 missing values and the `age` variable contains 12 missing values. Both `bmi` and `age` are registered as imputed. If we assume that age and BMI are independent, we can impute each of them separately by using the previously described univariate imputation methods. It is likely, however, that these variables are related, and so we use multivariate imputation.

First, we examine missing-value patterns of the data.

```
. mi misstable pattern

   Missing-value patterns
     (1 means complete)

              |  Pattern
      Percent |  1  2
              |
          82% |  1  1
              |
           10 |  1  0
            8 |  0  0
              |
         100% |

  Variables are  (1) age   (2) bmi
```

From the output, 82% of observations are complete, 10% of observations contain missing values for bmi, and 8% of observations have both bmi and age missing. We can see that the dataset has a monotone-missing pattern (see [MI] **intro substantive**), i.e., missing values of age are nested within missing values of bmi. Another way to see if the pattern of missingness is monotone is to use mi misstable nested ([MI] **mi misstable**):

```
. mi misstable nested
   1.   age(12) -> bmi(28)
```

Because the missing-data pattern is monotone, we can use mi impute monotone to impute missing values of bmi and age simultaneously:

```
. mi impute monotone (regress) age bmi = attack smokes hsgrad female, add(10)
Conditional models:
           age: regress age attack smokes hsgrad female
           bmi: regress bmi age attack smokes hsgrad female

Multivariate imputation               Imputations =       10
Monotone method                             added =       10
Imputed: m=1 through m=10                 updated =        0

           age: linear regression
           bmi: linear regression

                    |        Observations per m
                    |-----------------------------------
           Variable |  complete   incomplete   imputed  |   total
                    |                                   |
                age |       142           12        12  |     154
                bmi |       126           28        28  |     154

(complete + incomplete = total; imputed is the minimum across m
 of the number of filled in observations.)
```

Without going into detail, mi impute monotone imputes missing values of multiple variables by performing a sequence of independent univariate conditional imputations. In the above example, the regression method is used to impute missing values of both variables. age is imputed first from the observed variables attack, smokes, hsgrad, and female. Then bmi is imputed using the imputed age variable in addition to other observed variables. The output is consistent with that of the univariate imputation methods described earlier, with some additional information. See [MI] **mi impute monotone** for details.

Alternatively, we can use `mi impute mvn` to impute missing values of `bmi` and `age` simultaneously:

```
. mi impute mvn age bmi = attack smokes hsgrad female, replace nolog
Multivariate imputation              Imputations  =       10
Multivariate normal regression             added  =        0
Imputed: m=1 through m=10                updated  =       10
Prior: uniform                        Iterations  =     1000
                                         burn-in  =      100
                                         between  =      100
```

	Observations per m			
Variable	complete	incomplete	imputed	total
age	142	12	12	154
bmi	126	28	28	154

(complete + incomplete = total; imputed is the minimum across m
of the number of filled in observations.)

`mi impute mvn` does not require the missing-data pattern to be monotone. `mi impute mvn` iterates to produce imputations. When the data are monotone missing, however, no iteration is required and because `mi impute monotone` executes more quickly, it is preferred. Use `mi impute mvn` when there is an arbitrary missing-data pattern. See [MI] **mi impute mvn** for details.

Imputing on subsamples

Consider the earlier example of the univariate imputation of `bmi`. Suppose that we want to perform imputation separately for females and males. Imputation on subsamples is useful when the imputation model must accommodate the interaction effects (see, for example, Allison [2001]). For example, if we want the effect of `bmi` on `attack` to vary by gender, we can perform imputation of `bmi` separately for females and males:

```
. use http://www.stata-press.com/data/r11/mheart1s0
(Fictional heart attack data; bmi missing)
. mi impute regress bmi attack smokes age hsgrad if female==1, add(20)
Univariate imputation                Imputations  =       20
Linear regression                          added  =       20
Imputed: m=1 through m=20                updated  =        0
```

	Observations per m			
Variable	complete	incomplete	imputed	total
bmi	33	5	5	38

(complete + incomplete = total; imputed is the minimum across m
of the number of filled in observations.)

```
. mi impute regress bmi attack smokes age hsgrad if female==0, replace
Univariate imputation                Imputations  =       20
Linear regression                          added  =        0
Imputed: m=1 through m=20                updated  =       20
```

	Observations per m			
Variable	complete	incomplete	imputed	total
bmi	99	17	17	116

(complete + incomplete = total; imputed is the minimum across m
of the number of filled in observations.)

First, we created 20 imputations and filled in the missing values of bmi for females by using the corresponding subset of observations. Then we filled in the remaining missing values of bmi for males in the existing imputations by using the subset of male observations. We will then be able to include the interaction between bmi and female in our logistic model.

Imputation and estimation samples

Rubin (1987, 160–166) describes the imputation process as three tasks: modeling, estimation, and imputation. We concentrate on the latter two tasks here. The posterior distribution of the model parameters is estimated during the estimation task. This posterior distribution is used in the imputation task to simulate the parameters of the posterior predictive distribution of the missing data from which an imputed value is drawn. Accordingly, mi impute distinguishes between the two samples: imputation and estimation.

The imputation sample is determined by the imputation variables used in the imputation task. It is comprised of all observations for which the imputation variables contain nonmissing (more precisely, non–soft missing). In other words, the imputation sample consists of the complete and incomplete observations as identified by the specified imputation variables. The estimation sample is comprised of all observations used by the model fit to the observed data during the estimation task.

For example,

```
. use http://www.stata-press.com/data/r11/mheart1s0
(Fictional heart attack data; bmi missing)
. mi impute regress bmi attack smokes age hsgrad female, add(1) noisily

Running regress on observed data:
```

Source	SS	df	MS		Number of obs	=	132
					F(5, 126)	=	1.24
Model	99.5998228	5	19.9199646		Prob > F	=	0.2946
Residual	2024.93667	126	16.070926		R-squared	=	0.0469
					Adj R-squared	=	0.0091
Total	2124.5365	131	16.2178358		Root MSE	=	4.0089

bmi	Coef.	Std. Err.	t	P>|t|	[95% Conf. Interval]	
attack	1.71356	.7515229	2.28	0.024	.2263179	3.200801
smokes	-.5153181	.761685	-0.68	0.500	-2.02267	.9920341
age	-.033553	.0305745	-1.10	0.275	-.0940591	.026953
hsgrad	-.4674308	.8112327	-0.58	0.566	-2.072836	1.137975
female	-.3072767	.8074763	-0.38	0.704	-1.905249	1.290695
_cons	26.96559	1.884309	14.31	0.000	23.2366	30.69458

```
Univariate imputation            Imputations =        1
Linear regression                      added =        1
Imputed: m=1                         updated =        0
```

	Observations per m			
Variable	complete	incomplete	imputed	total
bmi	132	22	22	154

(complete + incomplete = total; imputed is the minimum across m of the number of filled in observations.)

The imputation sample contains 154 observations and the estimation sample contains 132 observations (from the regression output). The estimation task of mi impute regress consists of fitting a linear regression of bmi on other variables to the observed data. We specified the noisily option to see results from the estimation task. Usually, the number of complete observations in the imputation sample (132 in this example) will be equal to the number of observations used in the estimation. Sometimes, however, observations may be dropped from the estimation—for example, if independent variables contain missing values. In this case, the number of complete observations in the imputation sample and the number of observations used in the estimation will be different. You should evaluate such cases to verify that results are as expected.

In general, missing values in independent variables (or in a weighting variable) do not affect the imputation sample but they may lead to missing imputed values. For example, in the above example, if age contained missing values in incomplete observations of bmi, the linear prediction for those observations would have been missing and thus the resulting imputed values would have been missing, too.

Imputing transformations of imputed variables

Continuing with the univariate example above, say that we discover that the distribution of bmi is skewed to the right, and thus we decide to impute bmi on the logarithmic scale instead of the original one. We can do this by creating a new variable, lnbmi, and imputing it instead of bmi.

What we will do is create lnbmi, register it as imputed, impute it, and then create bmi as a passive variable based on the formula bmi = exp(lnbmi).

We need to be careful when we create lnbmi to get its missing values right. mi respects two kinds of missing values, called soft and hard missing. Soft missing values are missing values eligible for imputation. Hard missing values are missing values that are to remain missing even in the imputed data. Soft missing are recorded as ordinary missing (.), and hard missing are recorded as any of extended missing (.a–.z).

The issue here is that missing values could arise because of our application of the transform lnbmi = ln(bmi). In the case of the ln() transform, missing values will be created whenever bmi $\leq$ 0. Body mass index does not contain such values, but let's pretend it did. Here is what we would do:

1. Create lnbmi = ln(bmi).
2. Replace lnbmi to contain .z in observations for which lnbmi contains missing but bmi does not.
3. Register lnbmi as an imputed variable and impute it.
4. Create passive variable newbmi = exp(lnbmi).
5. Replace newbmi equal to bmi in observations for which newbmi is missing and bmi is not.

As we said, for lnbmi = ln(bmi) we need not do this because bmi > 0. In the bmi case, all we need to do is

1. Create lnbmi = ln(bmi).
2. Register lnbmi as an imputed variable and impute it.
3. Create passive variable newbmi = exp(lnbmi).

If all we wanted to do was impute lnbmi = ln(bmi) and, from that point on, just work with lnbmi, we would perform only the first two steps of the three-step procedure.

All that said, we are going to perform the five-step procedure because it will always work. We will continue from where we left off in the last example, so we will discard our previous imputation efforts by typing mi set M = 0. (Instead of typing mi set M = 0, we could just as easily begin by typing use http://www.stata-press.com/data/r11/mheart1s0.)

```
. mi set M = 0                              // start again
. mi unregister bmi                         // we do not impute bmi
. generate lnbmi = ln(bmi)                  // create lnbmi
. replace lnbmi = .z if lnbmi==. & bmi!=.
. mi register imputed lnbmi
. mi impute regress lnbmi attack smokes age hsgrad female, add(5)
. mi passive: gen newbmi = exp(bmi)
. mi passive: replace newbmi = bmi if bmi!=.
```

The important thing about the above is the very mechanical definition of an imputed variable. An imputed variable is a variable we actually impute, not a variable we desire to impute. In this case, we imputed lnbmi and derived bmi from it. Thus the variable we desired to impute became, mechanically, a passive variable.

Saved results

mi impute saves the following in r():

Scalars
- r(M)   total number of imputations
- r(M_add)   number of added imputations
- r(M_update)   number of updated imputations
- r(k_ivars)   number of imputed variables

Macros
- r(method)   name of imputation method
- r(ivars)   names of imputation variables
- r(rseed)   random-number seed

Matrices
- r(N)   number of observations in imputation sample (per variable)
- r(N_complete)   number of complete observations in imputation sample (per variable)
- r(N_incomplete)   number of incomplete observations in imputation sample (per variable)
- r(N_imputed)   number of imputed observations in imputation sample (per variable)

Also see *Saved results* in the method-specific manual entries for additional saved results.

Methods and formulas

All imputation methods are based on simulating from a Bayesian (approximate) posterior predictive distribution of missing data. Univariate imputation methods and the sequential monotone method use noniterative techniques for simulating from the posterior predictive distribution of missing data. The imputation method based on multivariate normal regression uses an iterative MCMC technique to simulate from the posterior predictive distribution of missing data.

Also see *Methods and formulas* in the method-specific manual entries for more details.

> Herman Otto Hartley (1912–1980) was born in Germany as Herman Otto Hirschfeld and immigrated to England in 1934 after completing his PhD in mathematics at Berlin University. He completed a second PhD in mathematical statistics under John Wishart at Cambridge in 1940 and went on to hold positions at Harper Adams Agricultural College, Scientific Computing Services (London), University College (London), Iowa State College, Texas A&M University, and Duke University. Among other awards he received and distinguished titles he held, Professor Hartley served as the president of the American Statistical Association in 1979. Known affectionately as HOH by almost all who knew him, he founded the Institute of Statistics, later to become the Department of Statistics, at Texas A&M University. His contributions to statistical computing are particularly notable considering the available equipment at the time. Professor Hartley is best known for his two-volume *Biometrika Tables for Statisticians* (jointly written with Egon Pearson) and for his fundamental contributions to sampling theory, missing-data methodology, variance-component estimation, and computational statistics.

References

Allison, P. D. 2001. *Missing Data*. Thousand Oaks, CA: Sage.

Graham, J. W. 2009. Missing data analysis: Making it work in the real world. *Annual Review of Psychology* 60: 549–576.

Raghunathan, T. E., J. M. Lepkowski, J. Van Hoewyk, and P. Solenberger. 2001. A multivariate technique for multiply imputing missing values using a sequence of regression models. *Survey Methodology* 27: 85–95.

Royston, P. 2007. Profile likelihood for estimation and confidence intervals. *Stata Journal* 7: 376–387.

Rubin, D. B. 1987. *Multiple Imputation for Nonresponse in Surveys*. New York: Wiley.

Schafer, J. L. 1997. *Analysis of Incomplete Multivariate Data*. Boca Raton, FL: Chapman & Hall/CRC.

Schenker, N., and J. M. G. Taylor. 1996. Partially parametric techniques for multiple imputation. *Computational Statistics & Data Analysis* 22: 425–446.

van Buuren, S. 2007. Multiple imputation of discrete and continuous data by fully conditional specification. *Statistical Methods in Medical Research* 16: 219–242.

Also see

[MI] **mi impute regress** — Impute using linear regression

[MI] **mi impute pmm** — Impute using predictive mean matching

[MI] **mi impute logit** — Impute using logistic regression

[MI] **mi impute mlogit** — Impute using multinomial logistic regression

[MI] **mi impute ologit** — Impute using ordered logistic regression

[MI] **mi impute monotone** — Impute missing values in monotone data

[MI] **mi impute mvn** — Impute using multivariate normal regression

[MI] **mi estimate** — Estimation using multiple imputations

[MI] **intro substantive** — Introduction to multiple-imputation analysis

[MI] **intro** — Introduction to mi

[MI] **Glossary**

Title

> **mi impute logit** — Impute using logistic regression

Syntax

mi impute logit *ivar* [*indepvars*] [*if*] [*weight*] [, *impute_options options*]

impute_options	description
Main	
*add(#)	specify number of imputations to add; required when no imputations exist
*replace	replace imputed values in existing imputations
rseed(#)	specify random-number seed
double	save imputed values in double precision; the default is to save them as float
Reporting	
dots	display dots as imputations are performed
noisily	display intermediate output
Advanced	
force	proceed with imputation even when missing imputed values are encountered
†noupdate	do not perform mi update; see [MI] **noupdate option**

*add(#) is required when no imputations exist; add(#) or replace is required if imputations exist.

†noupdate does not appear in the dialog box.

options	description
Main	
noconstant	suppress constant term
offset(*varname*)	include *varname* in model with coefficient constrained to 1
asis	retain perfect predictor variables in model
Maximization	
maximize_options	control the maximization process; seldom used

You must mi set your data before using mi impute logit; see [MI] **mi set**.
fweights, iweights, and pweights are allowed; see [U] **11.1.6 weight**.

Menu

Statistics > Multiple imputation

Description

mi impute logit fills in missing values of a binary variable by using a logistic regression imputation method. You can perform separate imputations on different subsets of the data by specifying an if condition. You can also account for frequency and sampling weights.

Options

┌─ Main ─┐

noconstant; see [R] **estimation options**.

add(), replace, rseed(), double; see [MI] **mi impute**.

offset(*varname*); see [R] **estimation options**.

asis forces retention of perfect predictor variables and their associated perfectly predicted observations in the model fit to the observed data. This may produce instabilities in maximization; see [R] **logit** and [R] **probit**.

┌─ Reporting ─┐

dots, noisily; see [MI] **mi impute**. noisily specifies that the output from the logistic regression fit to the observed data be displayed.

┌─ Maximization ─┐

maximize_options; see [R] **logit**. These options are seldom used.

┌─ Advanced ─┐

force; see [MI] **mi impute**.

noupdate; see [MI] **noupdate option**.

Remarks

Remarks are presented under the following headings:

> *Univariate imputation using logistic regression*
> *Using mi impute logit*

See [MI] **mi impute** for a general description and details about options common to all imputation methods. Also see [MI] **workflow** for general advice on working with mi.

Univariate imputation using logistic regression

The logistic regression imputation method can be used to fill in missing values of a binary variable (e.g., Rubin [1987]; Raghunathan et al. [2001]; and van Buuren [2007]). It is a parametric method that assumes an underlying logistic model for the imputed variable.

Unlike the linear regression method, the logistic imputation method is based on the asymptotic approximation of the posterior predictive distribution of the missing data. The actual posterior distribution of the logistic model parameters, β, does not have a simple form under the common noninformative prior distribution. Thus a large-sample normal approximation to the posterior distribution of β is used instead. Rubin (1987, 169) points out that although the actual posterior distribution may be far from normal (for example, when the number of observed cases is small or when the fraction of ones in the observed data is close to zero or one), the use of the normal approximation is common in practice.

Using mi impute logit

Continuing our heart attack example from [MI] **intro substantive** and [MI] **mi impute**, suppose that hsgrad, a binary variable recording whether subjects graduated from high school, contains missing values:

```
. use http://www.stata-press.com/data/r11/mheart2
(Fictional heart attack data; hsgrad missing)
. mi set mlong
. mi misstable summarize
```

Variable	Obs=.	Obs>.	Obs<.	Unique values	Obs<. Min	Max
hsgrad	18		136	2	0	1

Thus we want to impute missing values of hsgrad, because hsgrad was one of the predictors in our logistic model (logit attack smokes age bmi female hsgrad). From our previous analysis of the heart attack data, we recall that hsgrad was not a significant predictor. So, we could have omitted hsgrad from the logistic model in the casewise-deletion analysis to avoid the reduction in sample size, and then imputing hsgrad would not have been needed. In general, the imputer rarely has such knowledge, and omitting hsgrad from the imputation model would prevent this predictor from being used in later analysis by the analyst (see, e.g., *Imputation modeling* in [MI] **mi impute**). Thus we proceed with imputation.

We use mi impute logit to create 10 imputations of hsgrad:

```
. mi register imputed hsgrad
(18 m=0 obs. now marked as incomplete)
. mi impute logit hsgrad attack smokes age bmi female, add(10)
Univariate imputation              Imputations =       10
Logistic regression                      added =       10
Imputed: m=1 through m=10              updated =        0
```

Variable	Observations per m			
	complete	incomplete	imputed	total
hsgrad	136	18	18	154

(complete + incomplete = total; imputed is the minimum across m
of the number of filled in observations.)

We can now use the imputed hsgrad in our analysis, for example,

```
. mi estimate: logit attack smokes age bmi female hsgrad
  (output omitted)
```

Saved results

`mi impute logit` saves the following in `r()`:

Scalars
    `r(M)`   total number of imputations
    `r(M_add)`   number of added imputations
    `r(M_update)`   number of updated imputations
    `r(k_ivars)`   number of imputed variables (always one)

Macros
    `r(method)`   name of imputation method (`logit`)
    `r(ivars)`   names of imputation variables
    `r(rseed)`   random-number seed

Matrices
    `r(N)`   number of observations in imputation sample
    `r(N_complete)`   number of complete observations in imputation sample
    `r(N_incomplete)`   number of incomplete observations in imputation sample
    `r(N_imputed)`   number of imputed observations in imputation sample

Methods and formulas

Consider a univariate variable $\mathbf{x} = (x_1, x_2, \ldots, x_n)'$ that follows a logistic model

$$\Pr(x_i \neq 0 | \mathbf{z}_i) = \frac{\exp(\mathbf{z}'_i \boldsymbol{\beta})}{1 + \exp(\mathbf{z}'_i \boldsymbol{\beta})} \tag{1}$$

where $\mathbf{z}_i = (z_{i1}, z_{i2}, \ldots, z_{iq})'$ records values of predictors of $\mathbf{x}$ for observation i and $\boldsymbol{\beta}$ is the $q \times 1$ vector of unknown regression coefficients. (Note that when a constant is included in the model—the default—$z_{i1} = 1$, $i = 1, \ldots, n$.)

$\mathbf{x}$ contains missing values that are to be filled in. Consider the partition of $\mathbf{x} = (\mathbf{x}'_o, \mathbf{x}'_m)$ into $n_0 \times 1$ and $n_1 \times 1$ vectors containing the complete and the incomplete observations. Consider a similar partition of $\mathbf{Z} = (\mathbf{Z}_o, \mathbf{Z}_m)$ into $n_0 \times q$ and $n_1 \times q$ submatrices.

`mi impute logit` follows the steps below to fill in $\mathbf{x}_m$:

1. Fit a logistic model (1) to the observed data $(\mathbf{x}_o, \mathbf{Z}_o)$ to obtain the maximum likelihood estimates, $\widehat{\boldsymbol{\beta}}$, and their asymptotic sampling variance, $\widehat{\mathbf{U}}$.

2. Simulate new parameters, $\boldsymbol{\beta}_\star$, from the large-sample normal approximation, $N(\widehat{\boldsymbol{\beta}}, \widehat{\mathbf{U}})$, to its posterior distribution assuming the noninformative prior $\Pr(\boldsymbol{\beta}) \propto \text{const}$.

3. Obtain one set of imputed values, $\mathbf{x}^1_m$, by simulating from the logistic distribution:

$$\Pr(x_{i_m} = 1) = \exp(\mathbf{z}'_{i_m} \boldsymbol{\beta}_\star) / \left\{ 1 + \exp(\mathbf{z}'_{i_m} \boldsymbol{\beta}_\star) \right\}$$

for every missing observation i_m.

4. Repeat steps 2 and 3 to obtain M sets of imputed values, $\mathbf{x}^1_m, \mathbf{x}^2_m, \ldots, \mathbf{x}^M_m$.

Steps 2 and 3 above correspond to only approximate draws from the posterior predictive distribution of the missing data $\Pr(\mathbf{x}_m | \mathbf{x}_o, \mathbf{Z}_o)$ because $\boldsymbol{\beta}_\star$ is drawn from the asymptotic approximation to its posterior distribution.

If weights are specified, a weighted logistic regression model is fit to the observed data in step 1 (see [R] **logit** for details).

References

Raghunathan, T. E., J. M. Lepkowski, J. Van Hoewyk, and P. Solenberger. 2001. A multivariate technique for multiply imputing missing values using a sequence of regression models. *Survey Methodology* 27: 85–95.

Rubin, D. B. 1987. *Multiple Imputation for Nonresponse in Surveys.* New York: Wiley.

van Buuren, S. 2007. Multiple imputation of discrete and continuous data by fully conditional specification. *Statistical Methods in Medical Research* 16: 219–242.

Also see

[MI] **mi impute** — Impute missing values

[MI] **mi estimate** — Estimation using multiple imputations

[MI] **intro** — Introduction to mi

[MI] **intro substantive** — Introduction to multiple-imputation analysis

Title

> **mi impute mlogit** — Impute using multinomial logistic regression

Syntax

mi impute mlogit *ivar* [*indepvars*] [*if*] [*weight*] [, *impute_options options*]

impute_options	description
Main	
add(#)	specify number of imputations to add; required when no imputations exist
replace	replace imputed values in existing imputations
rseed(#)	specify random-number seed
double	save imputed values in double precision; the default is to save them as float
Reporting	
dots	display dots as imputations are performed
noisily	display intermediate output
Advanced	
force	proceed with imputation even when missing imputed values are encountered
†noupdate	do not perform mi update; see [MI] **noupdate option**

*add(#) is required when no imputations exist; add(#) or replace is required if imputations exist.

†noupdate does not appear in the dialog box.

options	description
Main	
noconstant	suppress constant term
baseoutcome(#)	specify value of *ivar* that will be the base outcome
Maximization	
maximize_options	control the maximization process; seldom used

You must mi set your data before using mi impute mlogit; see [MI] **mi set**.
fweights, iweights, and pweights are allowed; see [U] **11.1.6 weight**.

Menu

Statistics > Multiple imputation

mi impute mlogit — Impute using multinomial logistic regression

Description

mi impute mlogit fills in missing values of a nominal variable by using the multinomial (polytomous) logistic regression imputation method. You can perform separate imputations on different subsets of the data by specifying an if condition. You can also account for frequency and sampling weights.

Options

⌐ Main ⌐

noconstant; see [R] **estimation options**.

add(), replace, rseed(), double; see [MI] **mi impute**.

baseoutcome(#) specifies the value of *ivar* to be treated as the base outcome. The default is to choose the most frequent outcome.

⌐ Reporting ⌐

dots, noisily; see [MI] **mi impute**. noisily specifies that the output from the multinomial logistic regression fit to the observed data be displayed.

⌐ Maximization ⌐

maximize_options; see [R] **mlogit**. These options are seldom used.

⌐ Advanced ⌐

force; see [MI] **mi impute**.

noupdate; see [MI] **noupdate option**.

Remarks

Remarks are presented under the following headings:

>   *Univariate imputation using multinomial logistic regression*
>   *Using mi impute mlogit*

See [MI] **mi impute** for a general description and details about options common to all imputation methods. Also see [MI] **workflow** for general advice on working with mi.

Univariate imputation using multinomial logistic regression

The multinomial logistic regression imputation method can be used to fill in missing values of a nomial variable (e.g., Raghunathan et al. [2001] and van Buuren [2007]). It is a parametric method that assumes an underlying multinomial logistic model for the imputed variable. Similarly to the logistic imputation method, this method is based on the asymptotic approximation of the posterior predictive distribution of the missing data.

Using mi impute mlogit

Consider the heart attack data introduced in [MI] **intro substantive** and discussed in [MI] **mi impute**. Suppose that we want our logistic model of interest to also include information about marital status (categorical variable `marstatus`)—`logit attack smokes age bmi female hsgrad i.marstatus`.

We first tabulate values of `marstatus`:

```
. use http://www.stata-press.com/data/r11/mheart3
(Fictional heart attack data; marstatus missing)

. tabulate marstatus, missing
```

Marital status: single, married, divorced	Freq.	Percent	Cum.
Single	53	34.42	34.42
Married	48	31.17	65.58
Divorced	46	29.87	95.45
.	7	4.55	100.00
Total	154	100.00	

From the output, the `marstatus` variable has 3 unique categories and 7 missing observations. Because `marstatus` is a categorical variable, we use the multinomial logistic imputation method to fill in its missing values.

We `mi set` the data, register `marstatus` as an imputed variable, and then create 10 imputations by specifying the `add(10)` option with `mi impute mlogit`:

```
. mi set mlong
. mi register imputed marstatus
(7 m=0 obs. now marked as incomplete)
. mi impute mlogit marstatus attack smokes age bmi female hsgrad, add(10)
Univariate imputation              Imputations =      10
Multinomial logistic regression          added =      10
Imputed: m=1 through m=10              updated =       0
```

	Observations per m			
Variable	complete	incomplete	imputed	total
marstatus	147	7	7	154

(complete + incomplete = total; imputed is the minimum across m of the number of filled in observations.)

We can now analyze these multiply imputed data using logistic regression via `mi estimate`:

```
. mi estimate: logit attack smokes age bmi female hsgrad i.marstatus
  (output omitted)
```

Saved results

mi impute mlogit saves the following in r():

Scalars
    r(M)   total number of imputations
    r(M_add)   number of added imputations
    r(M_update)   number of updated imputations
    r(k_ivars)   number of imputed variables (always one)

Macros
    r(method)   name of imputation method (mlogit)
    r(ivars)   names of imputation variables
    r(rseed)   random-number seed

Matrices
    r(N)   number of observations in imputation sample
    r(N_complete)   number of complete observations in imputation sample
    r(N_incomplete)   number of incomplete observations in imputation sample
    r(N_imputed)   number of imputed observations in imputation sample

Methods and formulas

Consider a univariate variable $\mathbf{x} = (x_1, x_2, \ldots, x_n)'$ that contains K categories (without loss of generality, let $k = 1$ is the base outcome) and follows a multinomial logistic model

$$\Pr(x_i = k | \mathbf{z}_i) = \begin{cases} \dfrac{1}{1 + \sum_{l=2}^{K} \exp(\mathbf{z}_i' \boldsymbol{\beta}_l)}, & \text{if } k = 1 \\ \dfrac{\exp(\mathbf{z}_i' \boldsymbol{\beta}_k)}{1 + \sum_{l=2}^{K} \exp(\mathbf{z}_i' \boldsymbol{\beta}_l)}, & \text{if } k > 1 \end{cases} \quad (1)$$

where $\mathbf{z}_i = (z_{i1}, z_{i2}, \ldots, z_{iq})'$ records values of predictors of bx for observation i and $\boldsymbol{\beta}_l$ is the $q \times 1$ vector of unknown regression coefficients for outcome $l = 2, \ldots, K$. (Note that when a constant is included in the model—the default—$z_{i1} = 1$, $i = 1, \ldots, n$.)

$\mathbf{x}$ contains missing values that are to be filled in. Consider the partition of $\mathbf{x} = (\mathbf{x}_o', \mathbf{x}_m')$ into $n_0 \times 1$ and $n_1 \times 1$ vectors containing the complete and the incomplete observations. Consider a similar partition of $\mathbf{Z} = (\mathbf{Z}_o, \mathbf{Z}_m)$ into $n_0 \times q$ and $n_1 \times q$ submatrices.

mi impute mlogit follows the steps below to fill in $\mathbf{x}_m$:

1. Fit a multinomial logistic model (1) to the observed data $(\mathbf{x}_o, \mathbf{Z}_o)$ to obtain the maximum likelihood estimates, $\widehat{\boldsymbol{\beta}} = (\widehat{\boldsymbol{\beta}}_2', \ldots, \widehat{\boldsymbol{\beta}}_K')'$, and their asymptotic sampling variance, $\widehat{\mathbf{U}}$.

2. Simulate new parameters, $\boldsymbol{\beta}_\star$, from the large-sample normal approximation, $N(\widehat{\boldsymbol{\beta}}, \widehat{\mathbf{U}})$, to its posterior distribution assuming the noninformative prior $\Pr(\boldsymbol{\beta}) \propto \text{const}$.

3. Obtain one set of imputed values, $\mathbf{x}_m^1$, by simulating from the multinomial logistic distribution: one of K categories is randomly assigned to a missing category, i_m, using the cumulative probabilities computed from (1) with $\boldsymbol{\beta}_l = \boldsymbol{\beta}_{\star l}$ and $\mathbf{z}_i = \mathbf{z}_{i_m}$.

4. Repeat steps 2 and 3 to obtain M sets of imputed values, $\mathbf{x}_m^1, \mathbf{x}_m^2, \ldots, \mathbf{x}_m^M$.

Steps 2 and 3 above correspond to only approximate draws from the posterior predictive distribution of the missing data $\Pr(\mathbf{x}_m|\mathbf{x}_o, \mathbf{Z}_o)$ because $\beta_\star$ is drawn from the asymptotic approximation to its posterior distribution.

If weights are specified, a weighted multinomial logistic regression model is fit to the observed data in step 1 (see [R] **mlogit** for details).

References

Raghunathan, T. E., J. M. Lepkowski, J. Van Hoewyk, and P. Solenberger. 2001. A multivariate technique for multiply imputing missing values using a sequence of regression models. *Survey Methodology* 27: 85–95.

van Buuren, S. 2007. Multiple imputation of discrete and continuous data by fully conditional specification. *Statistical Methods in Medical Research* 16: 219–242.

Also see

[MI] **mi impute** — Impute missing values

[MI] **mi impute ologit** — Impute using ordered logistic regression

[MI] **mi estimate** — Estimation using multiple imputations

[MI] **intro** — Introduction to mi

[MI] **intro substantive** — Introduction to multiple-imputation analysis

Title

> **mi impute monotone** — Impute missing values in monotone data

Syntax

Full (default) specification of prediction equations, basic syntax

    mi impute monotone (*uvmethod*) *ivars* [= *indepvars*] [*if*] [*weight*] [, *impute_options options*]

Full (default) specification of prediction equations, full syntax

    mi impute monotone *lhs* [= *indepvars*] [*if*] [*weight*] [, *impute_options options*]

where *lhs* is *lhs_spec* [*lhs_spec* [...]] and *lhs_spec* is

    (*uvmethod* [*if*] [*weight*] [, *uv_options* noisily]) *ivars*

Custom specification of prediction equations

    mi impute monotone *cmodels* [*if*] [*weight*] , custom [*impute_options options*]

where *cmodels* is (*cond_spec*) [(*cond_spec*) [...]] and a conditional specification, *cond_spec*, is

    uvmethod ivar [*rhs_spec*] [*if*] [*weight*] [, *uv_options* noisily]

rhs_spec includes *indepvars* and expressions of imputation variables bound in parentheses.

uv_options are the method-specific *options* as described in the manual entry for each *uvmethod*.

uvmethod	description
regress	linear regression for a continuous variable; [MI] **mi impute regress**
pmm	predictive mean matching for a continuous variable; [MI] **mi impute pmm**
logit	logistic regression for a binary variable; [MI] **mi impute logit**
ologit	ordinal logistic regression for an ordinal variable; [MI] **mi impute ologit**
mlogit	multinomial logistic regression for a nominal variable; [MI] **mi impute mlogit**

130   mi impute monotone — Impute missing values in monotone data

impute_options	description
Main	
* add(#)	specify number of imputations to add; required when no imputations exist
* replace	replace imputed values in existing imputations
rseed(#)	specify random-number seed
double	save imputed values in double precision; the default is to save them as float
Reporting	
dots	display dots as imputations are performed
<u>noi</u>sily	display intermediate output
Advanced	
force	proceed with imputation even when missing imputed values are encountered
† noupdate	do not perform mi update; see [MI] **noupdate option**

*add(#) is required when no imputations exist; add(#) or replace is required if imputations exist.

†noupdate does not appear in the dialog box.

options	description
Main	
* <u>c</u>ustom	customize prediction equations of conditional specifications
Reporting	
dryrun	show conditional specifications without imputing data
verbose	show conditional specifications and impute data; implied when custom prediction equations are not specified
report	show report about each conditional specification
<u>noleg</u>end	do not display a legend showing conditional imputation methods used
Advanced	
nomonotonechk	do not check whether variables follow a monotone-missing pattern

*custom is required when specifying customized prediction equations.

You must mi set your data before using mi impute monotone; see [MI] **mi set**.

fweights, aweights (regress and pmm only), iweights, and pweights are allowed; see [U] **11.1.6 weight**.

Menu

Statistics > Multiple imputation

Description

mi impute monotone fills in missing values in multiple variables by using a sequence of independent univariate conditional imputation methods. Variables to be imputed, *ivars*, must follow a monotone-missing pattern ([MI] **intro substantive**). You can perform separate imputations on different subsets of the data by specifying an if condition. You can also account for frequency, analytic (with regression or predictive mean matching imputation only), and sampling weights.

Options

___Main___

custom is required to build customized prediction equations within the univariate conditional specifications. Otherwise, the default—full—specification of prediction equations is assumed.

add(), replace, rseed(), double; see [MI] **mi impute**.

___Reporting___

dots, noisily; see [MI] **mi impute**. noisily specifies that the output from all univariate conditional models fit to the observed data be displayed.

dryrun specifies to show the conditional specifications that would be used to impute each variable without actually imputing data.

verbose specifies to show conditional specifications and impute data. verbose is implied when custom prediction equations are not specified.

report specifies to show a report about each univariate conditional specification.

nolegend suppresses a legend with the titles of the conditional imputation methods used.

___Advanced___

force; see [MI] **mi impute**.

noupdate; see [MI] **noupdate option**.

nomonotonechk specifies not to check that imputation variables follow a monotone-missing pattern. This option may be used to avoid potentially time-consuming checks. The monotonicity check may be time consuming when a large number of variables is being imputed. If you use nomonotonechk with a custom specification, make sure that you list the univariate conditional specifications in the order of monotonicity or you might obtain incorrect results.

Remarks

Remarks are presented under the following headings:

> Multivariate imputation when a missing-data pattern is monotone
> First use
> Using mi impute monotone
> Default syntax of mi impute monotone
> The alternative syntax of mi impute monotone—custom prediction equations
> Examples of using default prediction equations
> Examples of using custom prediction equations

See [MI] **mi impute** for a general description and details about options, *impute_options*, common to all imputation methods. Also see [MI] **workflow** for general advice on working with mi.

Multivariate imputation when a missing-data pattern is monotone

When a pattern of missingness in multiple variables is monotone (or, more rigorously, when the missingness-modeling structure is monotone distinct), a multivariate imputation can be replaced with a set of conditional univariate imputations (Rubin 1987, 170–178). Let $X_1, X_2, \ldots, X_p$ be ordered such that if X_{1j} is missing, then X_{2j} is also missing, although X_2 may also be missing in other observations; if X_{2j} is missing, then X_{3j} is missing, although X_3 may also be missing in other observations; and so on. Then a simultaneous imputation of variables $X_1, X_2, \ldots, X_p$ according to a model, $f()$, and complete predictors (independent variables), Z, is equivalent to the sequential conditional imputation:

$$\begin{aligned} X_1^\star &\sim f_1(X_1|Z) \\ X_2^\star &\sim f_2(X_2|Z, X_1^\star) \\ &\ldots \\ X_p^\star &\sim f_p(X_p|Z, X_1^\star, X_2^\star, \ldots, X_{p-1}^\star) \end{aligned} \tag{1}$$

where for brevity we omit conditioning on the model parameters. The univariate conditional imputation models, $f_j()$, can each be of a different type (normal, logistic, etc.), as is appropriate for imputing X_j.

The specification of a conditional imputation model, $f_j()$, includes an imputation method and a prediction equation relating an imputation variable to other explanatory variables. In what follows, we distinguish between the default (full) specification in which the identities of the complete explanatory variables are the same for all imputed variables, and the custom specification in which the identities are allowed to differ.

Under the full specification, prediction equations of each imputation variable include all complete independent variables and all preceding imputation variables that have already been imputed. Under the custom specification, each prediction equation may include a subset of the predictors that would be used under the full specification. The custom specification implies nothing more than the assumption of independence between certain imputation variables and certain sets of predictors.

Model (1) corresponds to the full specification. For example, consider imputation variables X_1, X_2, X_3, ordered from the most observed to the least observed, and complete predictors Z_1, Z_2. Under the full specification, the individual prediction equations are determined as follows. The most observed variable, X_1, is predicted from Z_1 and Z_2. The next most observed variable, X_2, is predicted from Z_1, Z_2, and previously imputed X_1. The least observed variable, X_3, is predicted from Z_1, Z_2, and previously imputed X_1 and X_2. (A constant is included in all prediction equations, by default.) We use the following notation to refer to the above sequence of prediction equations (imputation sequence): $X_1|Z_1, Z_2 \to X_2|X_1, Z_1, Z_2 \to X_3|X_1, X_2, Z_1, Z_2$.

A sequence such as $X_1|Z_1 \to X_2|X_1, Z_1, Z_2 \to X_3|X_1, Z_2$ would correspond to a custom specification. Here X_1 is assumed to be independent of Z_2, and X_3 is assumed to be independent of Z_1.

The monotone-distinct structure offers much flexibility in building a multivariate imputation model. It simplifies the often intractable multivariate imputation task to a set of simpler univariate imputation tasks. In addition, it accommodates imputation of a mixture of types of variables. So, what's the catch? The catch is that the pattern of missingness is rarely monotone in practice. There are types of data for which a monotone-missing data pattern can occur naturally (for example, follow-up measurements). Usually, however, this happens only by chance.

There are several ways to proceed if your data are not monotone missing. You can discard the observations that violate the monotone-missing pattern, especially if there are very few such observations. You can assume independence among the sets of variables to create independent monotone patterns. For example, the missingness pattern for X_1, X_2, X_3, X_4, X_5 may not be monotone, but it may be for X_1, X_3 and for X_2, X_4, X_5. If it is reasonable to assume independence between these two sets of variables, you can then impute each set separately by using monotone imputation. Other alternatives are to use certain techniques to complete the missing-data pattern to monotone (see, e.g., Schafer 1997), to use an iterative sequential (full conditional) imputation

(Royston 2005, 2007; van Buuren, Boshuizen, and Knook 1999; Raghunathan et al. 2001), or to assume an explicit multivariate parametric model for the imputation variables (see [MI] **mi impute mvn**).

Throughout this entry, we will assume that the considered imputation variables are monotone missing.

First use

Before we describe various uses of `mi impute monotone`, let's look at an example.

Consider the heart attack data examining the relationship between heart attack and smoking. The `age` and `bmi` variables contain missing values and follow a monotone-missing pattern. Recall multivariate imputation of `bmi` and `age` using `mi impute monotone` described in *Multivariate imputation* of [MI] **mi impute**:

```
. use http://www.stata-press.com/data/r11/mheart5s0
(Fictional heart attack data; bmi and age missing)
. mi impute monotone (regress) bmi age = attack smokes hsgrad female, add(10)
Conditional models:
          age: regress age attack smokes hsgrad female
          bmi: regress bmi age attack smokes hsgrad female

Multivariate imputation                     Imputations =       10
Monotone method                                   added =       10
Imputed: m=1 through m=10                       updated =        0

          age: linear regression
          bmi: linear regression

                        Observations per m
         -----------------------------------------------
         Variable |   complete   incomplete   imputed  |     total
         ---------+-----------------------------------+----------
              age |        142           12        12  |       154
              bmi |        126           28        28  |       154
         -----------------------------------------------

(complete + incomplete = total; imputed is the minimum across m
 of the number of filled in observations.)
```

The `age` and `bmi` variables have monotone missingness, and so `mi impute monotone` is used to fill in missing values. Ten imputations are created (`add(10)` option). The linear regression imputation method (`regress`) is used to impute both continuous variables. The `attack`, `smokes`, `hsgrad`, and `female` variables are used as complete predictors (independent variables).

The conditional models legend shows that `age` (having the least number of missing values) is imputed first using the `regress` method, even though we specified `bmi` before `age` on the `mi impute` command. After that, `bmi` is imputed using the `regress` method and the previously imputed variable `age` and the other predictors.

The header and table output were described in detail in [MI] **mi impute**. The additional information below the table is the legend describing what univariate imputation method was used to impute each variable. (If desired, this legend may be suppressed by specifying the `nolegend` option.)

Using mi impute monotone

Below we summarize general capabilities of mi impute monotone.

1. mi impute monotone requires that the specified imputation variables follow a monotone-missing pattern. If they do not, it will stop with an error:

   ```
   . mi impute monotone x1 x2 ...
   x1 x2: not monotone;
       imputation variables must have a monotone-missing structure;
       see mi misstable nested
   r(459);
   ```

   As indicated by the error message, we can use mi misstable nested to verify for ourselves that the imputation variables are not monotone missing. We could also use other features of mi misstable to investigate the pattern.

2. mi impute monotone offers two main syntaxes—one using the default (full) prediction equations,

   ```
   . mi impute monotone ...
   ```

   and the other allowing customization of prediction equations,

   ```
   . mi impute monotone ..., custom ...
   ```

   We will refer to the two syntaxes as default and custom, respectively.

3. mi impute monotone allows specification of a global (outer) if condition,

   ```
   . mi impute monotone ... if exp ...
   ```

   and equation-specific (inner) if conditions,

   ```
   . mi impute monotone ... (... if exp ...) ...
   ```

   A global if is applied to all equations (conditional specifications). You may combine global and equation-specific if conditions:

   ```
   . mi impute monotone ... (... if exp ...) ... if exp ...
   ```

4. mi impute monotone allows specification of global weights,

   ```
   . mi impute monotone ... [weight] ...
   ```

   and equation-specific weights,

   ```
   . mi impute monotone ... (... [weight] ...) ...
   ```

   Global weights are applied to all conditional specifications. Unlike if conditions, global and equation-specific weights may not be combined.

In the next two sections, we describe the use of mi impute monotone first using hypothetical situations and then using real examples.

Default syntax of mi impute monotone

We showed in *First use* an example of mi impute monotone with default prediction equations using the heart attack data. Here we provide more details about this default specification.

By default, mi impute monotone imputes missing values by using the full specification. It builds the corresponding univariate conditional imputation models based on the supplied information: *uvmethod*, the imputation method; *ivars*, the imputation variables; and *indepvars*, the complete predictors or independent variables.

Suppose that continuous variables x1, x2, and x3 contain missing values with a monotone-missing pattern. We want to impute these variables, and we decide to use the same univariate imputation method, say, linear regression, for all. We can do this by typing

        . mi impute monotone (regress) x1 x2 x3 ...

The above corresponds to the first syntax diagram of mi impute monotone: *uvmethod* is regress and *ivars* is x1 x2 x3. Relating the above to the model notation used in (1), f_1, f_2, f_3 represent linear regression imputation models and the prediction sequence is $X_1 \to X_2|X_1 \to X_3|X_1, X_2$.

If we have additional covariates containing no missing values (say, z1 and z2) that we want to include in the imputation model, we can do it by typing

        . mi impute monotone (regress) x1 x2 x3 = z1 z2 ...

Now *indepvars* is z1 z2 and the prediction sequence is $X_1|Z_1, Z_2 \to X_2|X_1, Z_1, Z_2 \to X_3|X_1, X_2, Z_1, Z_2$. Independent variables are included in the prediction equations of all conditional models.

Suppose that we want to use a different imputation method for one of the variables—we want to impute x3 using predictive mean matching. We can do this as by typing

        . mi impute monotone (regress) x1 x2 (pmm) x3 = z1 z2 ...

The above corresponds to the second syntax diagram of mi impute monotone, a generalization of the first that accommodates differing imputation methods. The right-hand side of the equation is unchanged. z1 and z2 are included in all three prediction equations. The left-hand side now has two specifications: (regress) x1 x2 and (pmm) x3. In previous examples, we had only one left-hand-side specification, *lhs_spec*—(regress) x1 x2 x3. (Note that the number of left-hand-side specifications does not necessarily correspond to the number of conditional models; the latter is determined by the number of imputation variables.) In this example, x1 and x2 are imputed using linear regression, and x3 is imputed using predictive mean matching.

Now, instead of using the default one nearest neighbor with pmm, say that we want to use three, which requires pmm's knn(3) option. All method-specific options must be specified within the parentheses surrounding the method:

        . mi impute monotone (regress) x1 x2 (pmm, knn(3)) x3 = z1 z2 ...

Under the default specification, you can list imputation variables in any order and mi impute monotone will determine the correct ordering that follows the monotone-missing pattern.

Suppose now we want to restrict the imputation sample for x2 to observations where z1 is one. (We also omit pmm's knn() option here.) The corresponding syntax is

        . mi impute monotone (regress) x1 (regress if z1==1) x2 (pmm) x3 = z1 z2 ...

If, in addition to the above, we want to impute all variables using an overall subsample where z3 is one, we can specify the global if z3==1 condition:

        . mi impute monotone (regress) x1 (regress if z1==1) x2 (pmm) x3 = z1 z2
        > if z3==1 ...

When any imputation variable is categorical, mi impute monotone automatically includes it as a factor variable in the prediction equations of other imputation variables. In another example, suppose that x1 is a categorical variable and is imputed using the multinomial logistic method:

        . mi impute monotone (mlogit) x1 (regress) x2 x3 ...

The above will result in the prediction sequence $X_1 \to X_2|\text{i.}X_1 \to X_3|\text{i.}X_1, X_2$ where $\text{i.}X_1$ denotes the factors of X_1.

The alternative syntax of mi impute monotone—custom prediction equations

Consider the prediction sequence $X_1 \to X_2|X_1 \to X_3|X_1, X_2$. Suppose that we want to predict X_3 from X_1 rather than from X_1 and X_2. This could be achieved by simply imputing X_1, X_2 and X_1, X_3 separately because of the implied assumption that X_3 and X_2 are independent. However, with a larger number of variables and more complicated prediction rules, separate imputations may not be appealing. So customization of the prediction equations is a good alternative.

You customize prediction equations using the custom syntax (the third syntax) of mi impute monotone. You must specify the custom option to notify mi impute monotone that you are specifying custom prediction equations.

Under the custom syntax, you specify a separate conditional imputation model for each imputation variable. The specification of a conditional model is the same as that for the chosen univariate imputation method, but the entire model must be bound in parentheses, for example,

```
. mi impute monotone (regress x1)
                    (regress x2 x1)
                    (regress x3 x1)
                           , custom ...
```

Here we have three conditional specifications: (regress x1), (regress x2 x1), and (regress x3 x1). The corresponding prediction sequence is $X_1 \to X_2|X_1 \to X_3|X_1$. Prediction equations have the syntax *ivar* [*rhs_spec*].

When specifying custom prediction equations, you are required to list the conditional models in the correct order of missing monotonicity. mi impute monotone will issue an error if you are wrong:

```
mi impute monotone: incorrect equation order
    equations must be listed in the monotone-missing order of the imputation
    variables (from most observed to least observed);  x2(2) -> x1(5) -> x3(10)
r(198);
```

If we have additional covariates z1 and z2 containing no missing values, we can include them in the imputation model:

```
. mi impute monotone (regress x1 z1 z2)
                    (regress x2 x1 z1 z2)
                    (regress x3 x1 z1 z2), custom ...
```

To use the predictive mean matching method for x3, we simply change the method from regress to pmm in the last conditional specification:

```
. mi impute monotone (regress x1 z1 z2)
                    (regress x2 x1 z1 z2)
                    (pmm x3 x1 z1 z2), custom ...
```

To include more nearest neighbors in pmm, we specify the knn(3) option within the last conditional specification:

```
. mi impute monotone (regress x1 z1 z2)
                    (regress x2 x1 z1 z2)
                    (pmm x3 x1 z1 z2, knn(3)), custom ...
```

Under the custom syntax, you can also include expressions of previously imputed variables in prediction equations. For example, if you want to model x3 using main and squared effects of x1 (ignoring predictors z1 and z2), you can type

```
. mi impute monotone (regress x1)
                    (regress x2 x1)
                    (pmm x3 x1 (x1^2)), custom ...
```

Note that we bound the expression x1^2 in parentheses. Any expression may appear inside the parentheses.

Similar to the default specification, we can include equation-specific `if`s,

```
. mi impute monotone (regress x1)
                    (regress x2 x1 if z1==1)
                    (pmm x3 x1), custom ...
```

and we can specify a global `if`,

```
. mi impute monotone (regress x1 z1 z2)
                    (regress x2 x1 z2 if z1==1)
                    (pmm x3 x1 z1 z2)
                                    if z3==1, custom ...
```

Suppose that one of the imputed variables is categorical. We can use the multinomial logistic method to impute its values:

```
. mi impute monotone (mlogit x1)
                    (regress x2 i.x1)
                    (regress x3 i.x1)
                                    , custom ...
```

Examples of using default prediction equations

▷ Example 1: Different imputation methods

Recall the heart attack example from *First use*. If we wanted to impute age using predictive mean matching instead of linear regression, we could type

```
. use http://www.stata-press.com/data/r11/mheart5s0
(Fictional heart attack data; bmi and age missing)
. mi impute monotone (regress) bmi (pmm) age = attack smokes hsgrad female,
> add(10)
Conditional models:
            age: pmm age attack smokes hsgrad female
            bmi: regress bmi age attack smokes hsgrad female

Multivariate imputation              Imputations =       10
Monotone method                            added =       10
Imputed: m=1 through m=10                updated =        0

            age: predictive mean matching
            bmi: linear regression
```

	Observations per m	
Variable	complete	incomplete	imputed	total
age	142	12	12	154
bmi	126	28	28	154

```
(complete + incomplete = total; imputed is the minimum across m
 of the number of filled in observations.)
```

As previously, we listed age and bmi in the reverse order here, and `mi impute monotone` determined the correct order of missing monotonicity.

◁

▷ Example 2: Imputing a variable on a subsample

Consider an `mi set` version of the heart attack data containing the indicator for smoking high-tar cigarettes (variable `hightar`):

```
. use http://www.stata-press.com/data/r11/mheart6s0
(Fictional heart attack data; bmi, age, and hightar missing)
. mi describe
  Style:  mlong
          last mi update 19jun2009 10:50:18, 2 days ago
  Obs.:   complete           124
          incomplete          30    (M = 0 imputations)
          ─────────────────────
          total              154
  Vars.:  imputed:   3; bmi(24) age(30) hightar(8)
          passive:   0
          regular:   4; attack smokes female hsgrad
          system:    3; _mi_m _mi_id _mi_miss
          (there are no unregistered variables)
```

`mi describe` reports that there are no imputations, three registered imputed variables (`hightar` is one of them), and four registered regular variables.

Next we use `mi misstable nested` to examine missing-data patterns in the data.

```
. mi misstable nested
    1.  hightar(8) -> bmi(24) -> age(30)
```

There is one monotone-missing pattern in the data. According to the output, missing values of `hightar` are nested within `bmi` whose missing values are nested within `age`, so `hightar`, `bmi`, and `age` follow a monotone-missing pattern.

As before, to impute missing values of `age` and `bmi`, we use the regression method. The `hightar` variable is a binary variable, so we choose the logistic method to fill in its values (see [MI] **mi impute logit**). Because `hightar` records whether a subject smokes high-tar cigarettes, we use only those who smoke to impute its missing values. (If there were any missing values of `hightar` for the subjects who do not smoke, we would have replaced them with zeros.)

```
. mi impute monotone (reg) age bmi (logit if smokes) hightar
> = attack smokes hsgrad female, add(10)
Conditional models:
        hightar: logit hightar attack smokes hsgrad female if (smokes)
           bmi: regress bmi i.hightar attack smokes hsgrad female
           age: regress age bmi i.hightar attack smokes hsgrad female

Multivariate imputation               Imputations =       10
Monotone method                             added =       10
Imputed: m=1 through m=10                 updated =        0

        hightar: logistic regression
           bmi: linear regression
           age: linear regression
```

	Observations per m			
Variable	complete	incomplete	imputed	total
hightar	56	8	8	64
bmi	130	24	24	154
age	124	30	30	154

```
(complete + incomplete = total; imputed is the minimum across m
of the number of filled in observations.)
```

mi impute monotone reports which univariate conditional model was used to impute each variable. Because hightar has the least number of missing observations, it is imputed first using the specified complete predictors and using only observations for smokers. From the output, all incomplete values of each of the variables are imputed in all 10 imputations. Notice that because we restricted the imputation sample of hightar to smokers, the total number of observations reported for hightar is 64 and not 154.

◁

Examples of using custom prediction equations

▷ Example 3: Using different sets of predictors within individual conditional models

Let's take a closer look at the conditional model for hightar used in the above example:

```
hightar: logit hightar attack smokes hsgrad female if (smokes)
```

Notice that predictor smokes is redundant in this model because it is collinear with the constant (included in the model by default) on the restricted sample of smokers. In fact, if we specify the noisily option (noi for short) within the logit specification to see the estimation results, we will notice that, as expected, smokes was omitted from the estimation model; that is, its coefficient is zero.

(Continued on next page)

```
. mi impute monotone (reg) age bmi (logit if smokes, noi) hightar
> = attack smokes hsgrad female, replace
Conditional models:
     hightar: logit hightar attack smokes hsgrad female if (smokes) , noi
        bmi: regress bmi i.hightar attack smokes hsgrad female
        age: regress age bmi i.hightar attack smokes hsgrad female

Running logit on observed data:

note: smokes omitted because of collinearity

Iteration 0:    log likelihood = -38.673263
Iteration 1:    log likelihood = -38.455029
Iteration 2:    log likelihood = -38.454991
Iteration 3:    log likelihood = -38.454991

Logistic regression                               Number of obs    =        56
                                                  LR chi2(3)       =      0.44
                                                  Prob > chi2      =    0.9326
Log likelihood = -38.454991                       Pseudo R2        =    0.0056
```

hightar	Coef.	Std. Err.	z	P>|z|	[95% Conf.	Interval]
attack	.0773715	.5630513	0.14	0.891	-1.026189	1.180932
hsgrad	-.1663937	.5977995	-0.28	0.781	-1.338059	1.005272
female	-.3331926	.617736	-0.54	0.590	-1.543933	.8775477
_cons	.0138334	.6263152	0.02	0.982	-1.213722	1.241389

```
Multivariate imputation                Imputations =      10
Monotone method                              added =       0
Imputed: m=1 through m=10                  updated =      10

     hightar: logistic regression
        bmi: linear regression
        age: linear regression
```

	Observations per m			
Variable	complete	incomplete	imputed	total
hightar	56	8	8	64
bmi	130	24	24	154
age	124	30	30	154

(complete + incomplete = total; imputed is the minimum across m
of the number of filled in observations.)

Although mi impute handles collinearity problems for us automatically, we can eliminate redundancy manually by removing smokes from the prediction equation for hightar. To do that, we need to specify custom prediction equations.

As discussed in *Using mi impute monotone*, custom prediction equations are available with mi impute monotone when the custom option is used. We also know that within this custom specification, we must fully specify prediction equations within each conditional model and must specify the conditional models in the monotone-missing order of the imputation variables.

Building such conditional models from scratch could be a tedious task except that we can use mi impute monotone, dryrun to display the conditional models with default prediction equations without performing the corresponding imputation:

```
. mi impute monotone (reg) age bmi (logit if smokes) hightar
>   = attack smokes hsgrad female, dryrun
Conditional models:
        hightar: logit hightar attack smokes hsgrad female if (smokes)
            bmi: regress bmi i.hightar attack smokes hsgrad female
            age: regress age bmi i.hightar attack smokes hsgrad female
```

We can use these default conditional specifications as the basis for writing our own customized specifications. We will remove smokes from the predictor list for hightar:

```
. mi impute monotone (logit hightar attack hsgrad female if smokes)
>                   (regress bmi hightar attack smokes hsgrad female)
>                   (regress age bmi hightar attack smokes hsgrad female)
>                                                           , custom replace
Multivariate imputation                    Imputations =       10
Monotone method                                  added =        0
Imputed: m=1 through m=10                      updated =       10

        hightar: logistic regression
            bmi: linear regression
            age: linear regression
```

	Observations per m			
Variable	complete	incomplete	imputed	total
hightar	56	8	8	64
bmi	130	24	24	154
age	124	30	30	154

(complete + incomplete = total; imputed is the minimum across m of the number of filled in observations.)

◁

▷ Example 4: Including expressions of imputation variables in prediction equations

The distribution of bmi is slightly skewed. To take this into account, we can either use predictive mean matching to impute bmi or impute bmi on a logarithmic scale. We choose to impute the log of bmi here.

Following the steps described in *Imputing transformations of imputed variables* of [MI] **mi impute**, we create a new variable, lnbmi, containing the log of bmi and register it as imputed. Here we also reset the number of imputations to zero.

```
. mi set M = 0
(10 imputations dropped; M = 0)
. mi unregister bmi
. generate lnbmi = ln(bmi)
(24 missing values generated)
. mi register imputed lnbmi
```

We are now ready to impute lnbmi. However, although we are imputing the log of bmi, we want to use bmi in the original scale when imputing age. To do that, we include exp(lnbmi) in the prediction equation for age. When including expressions in a custom specification, the expressions must appear in parentheses:

```
. mi impute monotone (logit hightar attack hsgrad female if smokes)
>                    (regress lnbmi hightar attack smokes hsgrad female)
>                    (regress age (exp(lnbmi)) hightar attack smokes hsgrad female)
>                                                                , custom add(10)
Multivariate imputation              Imputations =       10
Monotone method                            added =       10
Imputed: m=1 through m=10                updated =        0

    hightar: logistic regression
      lnbmi: linear regression
        age: linear regression
```

	Observations per m			
Variable	complete	incomplete	imputed	total
hightar	56	8	8	64
lnbmi	130	24	24	154
age	124	30	30	154

(complete + incomplete = total; imputed is the minimum across m of the number of filled in observations.)

If we also wanted to include a squared term for bmi in the conditional imputation model for age, we would type

```
. mi impute monotone
>    (logit hightar attack hsgrad female if smokes)
>    (regress lnbmi hightar attack smokes hsgrad female)
>    (regress age (exp(lnbmi)) (exp(lnbmi)^2) hightar attack smokes hsgrad female)
>                                                                , custom replace
  (output omitted)
```

◁

Saved results

mi impute monotone saves the following in r():

Scalars
    r(M)   total number of imputations
    r(M_add)   number of added imputations
    r(M_update)   number of updated imputations
    r(k_ivars)   number of imputed variables

Macros
    r(method)   name of imputation method (monotone)
    r(ivars)   names of imputation variables
    r(rseed)   random-number seed
    r(uvmethods)   names of univariate conditional imputation methods

Matrices
    r(N)   number of observations in imputation sample (per variable)
    r(N_complete)   number of complete observations in imputation sample (per variable)
    r(N_incomplete)   number of incomplete observations in imputation sample (per variable)
    r(N_imputed)   number of imputed observations in imputation sample (per variable)

Methods and formulas

Let $\mathbf{x}_{(i)} = (x_{i1}, x_{i2}, \ldots, x_{ip})$ be the ith observation containing values of the imputation variables ordered from the most observed to the least observed to form a monotone-missing data pattern. Let $\mathbf{z}_{(i)} = (z_{i1}, z_{i2}, \ldots, z_{iq})$ be the corresponding set of predictors of $\mathbf{x}_{(i)}$. Then, if the missingness-modeling structure is monotone distinct (imputation variables have monotone missingness and parameters of the conditional models are distinct as defined in Rubin [1987, 174]), the following decomposition holds:

$$f(\mathbf{x}_{(i)}|\mathbf{z}_{(i)}, \boldsymbol{\theta}) = f_1(x_{i1}|\mathbf{z}_{(i)}, \boldsymbol{\theta}_1) f_2(x_{i2}|\mathbf{z}_{(i)}, x_{i1}, \boldsymbol{\theta}_2) \cdots f(x_{ip}|\mathbf{z}_{(i)}, x_{i1}, x_{i2}, \ldots, x_{i,p-1}, \boldsymbol{\theta}_p)$$

where the unknown parameters $\boldsymbol{\theta}_1, \ldots, \boldsymbol{\theta}_p$ are distinct, that is, $\Pr(\boldsymbol{\theta}) = \prod_{j=1}^{p} \Pr(\boldsymbol{\theta}_j)$. The monotone-distinct structure ensures that univariate conditional models f_j do not depend on any unobserved values of variable $\mathbf{x}_j$ and the posterior distributions of $\boldsymbol{\theta}_j$ do not involve the imputed values of the previously filled in variables $\mathbf{x}_1, \ldots, \mathbf{x}_{j-1}$. See Rubin (1987, 174–178) for a rigorous justification of the above decomposition.

The above allows substituting the imputation of $\mathbf{X}$ using the probability model $f()$ with a sequence of univariate conditional imputations of $\mathbf{x}_j$ using the probability models $f_j()$. Note that f_j can be any proper imputation model (e.g., linear regression or logistic regression).

mi impute monotone follows the steps below to fill in missing values in $\mathbf{x}_1, \ldots, \mathbf{x}_p$:

1. If the custom option is not used, mi impute monotone first builds univariate conditional models containing the default (full) prediction equations using the supplied information about imputation methods, imputation variables $\mathbf{X}$, and complete predictors $\mathbf{Z}$. The order in which imputation variables are listed is irrelevant.

   If the custom option is used, mi impute monotone uses the specified conditional models in the order supplied. The conditional models must be listed in the monotone-missing order of the corresponding imputation variables.

2. Fit univariate conditional models for each $\mathbf{x}_i$ to the observed data to obtain the estimates of $\boldsymbol{\theta}_i$, $i = 1, \ldots, p$. See step 1 in *Methods and formulas* of each respective univariate imputation method's manual entry for details.

3. Sequentially fill in missing values of $\mathbf{x}_1, \mathbf{x}_2, \ldots, \mathbf{x}_p$ according to the specified imputation model. See step 2 and step 3 in *Methods and formulas* of each respective univariate imputation method's manual entry for details.

4. Repeat step 3 to obtain M multiple imputations.

References

Raghunathan, T. E., J. M. Lepkowski, J. Van Hoewyk, and P. Solenberger. 2001. A multivariate technique for multiply imputing missing values using a sequence of regression models. *Survey Methodology* 27: 85–95.

Royston, P. 2005. Multiple imputation of missing values: Update. *Stata Journal* 5: 188–201.

———. 2007. Profile likelihood for estimation and confidence intervals. *Stata Journal* 7: 376–387.

Rubin, D. B. 1987. *Multiple Imputation for Nonresponse in Surveys*. New York: Wiley.

Schafer, J. L. 1997. *Analysis of Incomplete Multivariate Data*. Boca Raton, FL: Chapman & Hall/CRC.

van Buuren, S., H. C. Boshuizen, and D. L. Knook. 1999. Multiple imputation of missing blood pressure covariates in survival analysis. *Statistics in Medicine* 18: 681–694.

Also see

[MI] **mi impute** — Impute missing values

[MI] **mi impute mvn** — Impute using multivariate normal regression

[MI] **mi estimate** — Estimation using multiple imputations

[MI] **intro substantive** — Introduction to multiple-imputation analysis

[MI] **intro** — Introduction to mi

[MI] **Glossary**

Title

> **mi impute mvn** — Impute using multivariate normal regression

Syntax

mi impute mvn *ivars* [= *indepvars*] [*if*] [, *impute_options options*]

impute_options	description
Main	
*add(#)	specify number of imputations to add; required when no imputations exist
*replace	replace imputed values in existing imputations
rseed(#)	specify random-number seed
double	save imputed values in double precision; the default is to save them as float
Reporting	
dots	display dots as imputations are performed
noisily	display intermediate output
Advanced	
force	proceed with imputation even when missing imputed values are encountered
†noupdate	do not perform mi update; see [MI] **noupdate option**

*add(#) is required when no imputations exist; add(#) or replace is required if imputations exist.

†noupdate does not appear in the dialog box.

options	description
Main	
noconstant	suppress constant term
MCMC options	
burnin(#)	specify number of iterations for the burn-in period; default is burnin(100)
burnbetween(#)	specify number of iterations between imputations; default is burnbetween(100)
prior(*prior_spec*)	specify a prior distribution; default is prior(uniform)
mcmconly	perform MCMC for the length of the burn-in period without imputing missing values
initmcmc(*init_mcmc*)	specify initial values for the MCMC procedure; default is initmcmc(em) using the EM estimates for initial values
wlfwgt(*matname*)	specify weights for the worst linear function
savewlf(*filename*[, ...])	save the worst linear function from each iteration in *filename*.dta
saveptrace(*fname*[, ...])	save MCMC parameter estimates from each iteration in *fname*.stptrace; see [MI] **mi ptrace**

Reporting

emlog	display iteration log from EM
emoutput	display intermediate output from EM estimation
alldots	display dots as imputation iterations are performed
nolog	do not display information about the EM or MCMC procedures

Advanced

emonly[(em_options)]	perform EM estimation only

indepvars may contain factor variables; see [U] **11.4.3 Factor variables**.
You must mi set your data before using mi impute mvn; see [MI] **mi set**.

prior_spec	description
<u>u</u>niform	use the uniform prior distribution; the default
<u>j</u>effreys	use the Jeffreys noninformative prior distribution
<u>r</u>idge, df(#)	use a ridge prior distribution with degrees of freedom #

init_mcmc	description
em[, *em_options*]	use EM to obtain starting values for MCMC; the default
initmatlist	supply matrices containing initial values for MCMC

em_options	description
<u>ite</u>rate(#)	specify the maximum number of iterations; default is iterate(100)
<u>tol</u>erance(#)	specify tolerance for the changes in parameter estimates; default is tolerance(1e-5)
init(*init_em*)	specify initial values for the EM algorithm; default is init(ac)
nolog	do not show EM iteration log
<u>savep</u>trace(*fname*[, ...])	save EM parameter estimates from each iteration in *fname*.stptrace; see [MI] **mi ptrace**

init_em	description
ac	use all available cases to obtain initial values for EM; the default
cc	use only complete cases to obtain initial values for EM
initmatlist	supply matrices containing initial values for EM

initmatlist is of the form *initmat* [*initmat* [...]]

initmat	description
<u>b</u>etas(# | *matname*)	specify coefficient vector; default is betas(0)
<u>sd</u>s(# | *matname*)	specify standard deviation vector; default is sds(1)
<u>var</u>s(# | *matname*)	specify variance vector; default is vars(1)
corr(# | *matname*)	specify correlation matrix; default is corr(0)
cov(*matname*)	specify covariance matrix

In the above, # is understood to mean a vector containing all elements equal to #.

Menu

Statistics > Multiple imputation

Description

mi impute mvn fills in missing values of one or more continuous variables using multivariate normal regression. It accommodates arbitrary missing-value patterns. You can perform separate imputations on different subsets of the data by specifying an if condition. mi impute mvn uses an iterative Markov chain Monte Carlo (MCMC) method to impute missing values. See *Remarks* for details.

Options

Main

noconstant; see [R] **estimation options**.

add(), replace, rseed(), double; see [MI] **mi impute**.

MCMC options

burnin(#) specifies the number of iterations for the initial burn-in period. The default is burnin(100). This option specifies the number of iterations necessary for the MCMC to reach approximate stationarity or, equivalently, to converge to a stationary distribution. The required length of the burn-in period will depend on the starting values used and the missing-data patterns observed in the data. It is important to examine the chain for convergence to determine an adequate length of the burn-in period prior to obtaining imputations; see *Convergence of the MCMC method* and examples 2 and 4. The provided default may be sufficient in many cases, but you are responsible for determining that sufficient iterations are performed.

burnbetween(#) specifies a number of iterations of the MCMC to perform between imputations, the purpose being to reduce correlation between sets of imputed values. The default is burnbetween(100). As with burnin(), you are responsible for determining that sufficient iterations are performed. See *Convergence of the MCMC method* and examples 2 and 4.

prior(*prior_spec*) specifies a prior distribution to be used by the MCMC procedure. The default is prior(uniform). The alternative prior distributions are useful when the default estimation of the parameters using maximum likelihood becomes unstable (for example, estimates on the boundary of the parameter space) and introducing some prior information about parameters stabilizes the estimation.

prior_spec is

uniform | jeffreys | ridge, df(#)

uniform specifies the uniform (flat) prior distribution. Under this prior distribution, the posterior distribution is proportional to the likelihood function and thus the estimate of the posterior mode is the same as the maximum likelihood (ML) estimate.

jeffreys specifies the Jeffreys, noninformative prior distribution. This prior distribution can be used when there is no strong prior knowledge about the model parameters.

ridge, df(#) specifies a ridge, informative prior distribution with the degrees of freedom #. This prior introduces some information about the covariance matrix by smoothing the off-diagonal elements (correlations) toward zero. The degrees of freedom, df(), which may be noninteger, regulate the amount of smoothness—the larger this number, the closer the correlations are to zero. A ridge prior is useful to stabilize inferences about the mean parameters when the covariance matrix is poorly estimated, for example, when there are insufficient observations to estimate correlations between some variables reliably because of missing data, causing the estimated covariance matrix to become nonpositive definite (see Schafer [1997, 155–157] for details).

mcmconly specifies that mi impute mvn run the MCMC for the length of the burn-in period and then stop. This option is useful in combination with savewlf() or saveptrace() to examine the convergence of the MCMC prior to imputation. No imputation is performed when mcmconly is specified, so add() or replace is not required with mi impute mvn, mcmconly and they are ignored if specified. The mcmconly option is not allowed with emonly.

initmcmc() may be specified as initmcmc(em [, em_options]) or initmcmc(initmatlist).

initmcmc() specifies initial values for the regression coefficients and covariance matrix of the multivariate normal distribution to be used by the MCMC procedure. By default, initial values are obtained from the EM algorithm, initmcmc(em).

initmcmc(em[, em_options]) specifies that the initial values for the MCMC procedure be obtained from EM. You can control the EM estimation by specifying em_options. If the uniform prior is used, the initial estimates correspond to the ML estimates computed using EM. Otherwise, the initial values are the estimates of the posterior mode computed using EM.

em_options are

iterate(#) specifies the maximum number of EM iterations to perform. The default is iterate(100).

tolerance(#) specifies the convergence tolerance for the EM algorithm. The default is tolerance(1e-5). Convergence is declared once the maximum of the relative changes between two successive estimates of all model parameters is less than #.

init() may be specified as init(ac), init(cc), or init(matlist)

init() specifies initial values for the regression coefficients and covariance matrix of the multivariate normal distribution to be used by the EM algorithm. init(ac) is the default.

init(ac) specifies that initial estimates be obtained using all available cases. The initial values for regression coefficients are obtained from separate univariate regressions of each imputation variable on the independent variables. The corresponding estimates of the residual mean-squared error are used as the initial values for the diagonal entries of the covariance matrix (variances). The off-diagonal entries (correlations) are set to zero.

init(cc) specifies that initial estimates be obtained using only complete cases. The initial values for regression coefficients and the covariance matrix are obtained from a multivariate regression fit to the complete cases only.

init(initmatlist) specifies to use manually supplied initial values for the EM procedure and syntactically is identical to mcmcinit(initmatlist), described below, except that you specify init(initmatlist).

nolog suppresses the EM iteration log when emonly or emoutput is used.

saveptrace(fname[, replace]) specifies to save the parameter trace log from the EM algorithm to a file called fname.stptrace. If the file already exists, the replace suboption specifies to overwrite the existing file. See [MI] **mi ptrace** for details about the saved file and how to read it into Stata.

initmcmc(*initmatlist*), where *initmatlist* is

initmat [*initmat* [...]]

specifies manually supplied initial values for the MCMC procedure.

initmat is

betas(#|*matname*) specifies initial values for the regression coefficients. The default is betas(0), implying a value of zero for all regression coefficients. If you specify betas(#), then # will be used as the initial value for all regression coefficients. Alternatively, you can specify the name of a Stata matrix, *matname*, containing values for each regression coefficient. *matname* must be conformable with the dimensionality of the specified model. That is, it can be one of the following dimensions: $p \times q$, $q \times p$, $1 \times pq$, or $pq \times 1$, where p is the number of imputation variables and q is the number of independent variables.

sds(#|*matname*) specifies initial values for the standard deviations (square roots of the diagonal elements of the covariance matrix). The default is sds(1), which sets all standard deviations and thus variances to one. If you specify sds(#), then the squared # will be used as the initial value for all variances. Alternatively, you can specify the name of a Stata matrix, *matname*, containing individual values. *matname* must be conformable with the dimensionality of the specified model. That is, it can be one of the following dimensions: $1 \times p$ or $p \times 1$, where p is the number of imputation variables. This option cannot be combined with cov() or vars(). The sds() option can be used in combination with corr() to provide initial values for the covariance matrix.

vars(#|*matname*) specifies initial values for variances (diagonal elements of the covariance matrix). The default is vars(1), which sets all variances to one. If you specify vars(#), then # will be used as the initial value for all variances. Alternatively, you can specify the name of a Stata matrix, *matname*, containing individual values. *matname* must be conformable with the dimensionality of the specified model. That is, it can be one of the following dimensions: $1 \times p$ or $p \times 1$, where p is the number of imputation variables. This option cannot be combined with cov() or sds(). The vars() option can be used in combination with corr() to provide initial values for the covariance matrix.

corr(#|*matname*) specifies initial values for the correlations (off-diagonal elements of the correlation matrix). The default is corr(0), which sets all correlations and, thus, covariances to zero. If you specify corr(#), then all correlation coefficients will be set to #. Alternatively, you can specify the name of a Stata matrix, *matname*, containing individual values. *matname* can be a square $p \times p$ matrix with diagonal elements equal to one or it can contain the corresponding lower (upper) triangular matrix in a vector of dimension $p(p+1)/2$, where p is the number of imputation variables. This option cannot be combined with cov(). The corr() option can be used in combination with sds() or vars() to provide initial values for the covariance matrix.

cov(*matname*) specifies initial values for the covariance matrix. *matname* must contain the name of a Stata matrix. *matname* can be a square $p \times p$ matrix or it can contain the corresponding lower (upper) triangular matrix in a vector of dimension $p(p+1)/2$, where p is the number of imputation variables. This option cannot be combined with corr(), sds(), or vars().

wlfwgt(*matname*) specifies the weights (coefficients) to use when computing the worst linear function (WLF). The coefficients must be saved in a Stata matrix, *matname*, of dimension $1 \times d$, where $d = pq + p(p+1)/2$, p is the number of imputation variables, and q is the number of predictors.

This option is useful when initial values from the EM estimation are supplied to data augmentation (DA) as matrices. This option can also be used to obtain the estimates of linear functions other than the default WLF.

savewlf(*filename*[, replace]) specifies to save the estimates of the WLF from each iteration of MCMC to a Stata dataset called *filename*.dta. If the file already exists, the replace suboption specifies to overwrite the existing file. This option is useful for monitoring convergence of the MCMC. savewlf() is allowed with initmcmc(em), when the initial values are obtained using the EM estimation, or with wlfwgt().

saveptrace(*fname*[, replace]) specifies to save the parameter trace log from the MCMC to a file called *fname*.stptrace. If the file already exists, the replace suboption specifies to overwrite the existing file. See [MI] **mi ptrace** for details about the saved file and how to read it into Stata. This option is useful for monitoring convergence of the MCMC.

⌐ Reporting ⌐

dots, noisily; see [MI] **mi impute**. Also, noisily is a synonym for emoutput.

emlog specifies that the EM iteration log be shown. The EM iteration log is not displayed unless emonly or emoutput is specified.

emoutput specifies that the EM output be shown. This option is implied with emonly.

alldots specifies to display all intermediate iterations as dots in addition to the imputation dots. The iterations include the EM iterations and the MCMC burn-in iterations. Also see the dots option in [MI] **mi impute**.

nolog suppresses all output from EM or MCMC that is usually displayed by default.

⌐ Advanced ⌐

force; see [MI] **mi impute**.

noupdate; see [MI] **noupdate option**.

emonly[(*em_options*)] specifies that mi impute mvn perform EM estimation and then stop. You can control the EM process by specifying *em_options*. This option is useful at the preliminary stage to obtain insight about the length of the burn-in period as well as to choose a prior specification. No imputation is performed, so add() or replace is not required with mi impute mvn, emonly, and they are ignored if specified. The emonly option is not allowed with mcmconly.

Remarks

Remarks are presented under the following headings:

> *Incomplete continuous data with arbitrary pattern of missing values*
> *Multivariate imputation using data augmentation*
> *Convergence of the MCMC method*
> *Using mi impute mvn*
> *Examples*

See [MI] **mi impute** for a general description and details about options common to all imputation methods. Also see [MI] **workflow** for general advice on working with mi.

Incomplete continuous data with arbitrary pattern of missing values

When the pattern of missing data is monotone, missing values can be filled in sequentially without iteration using univariate conditional imputation models, as discussed in [MI] **mi impute monotone**. This is possible because variables can be ordered such that the complete observations of a variable being imputed are also complete in all prior imputed variables used to predict it. This means that the estimates of the parameters, which are obtained from complete data, do not depend on any previously imputed values (see Rubin [1987] for details).

With an arbitrary pattern of missing data, such an ordering may not be possible; see *Patterns of missing data* in [MI] **intro substantive**. Some variables may contain incomplete values in observations for which other variables are complete and vice versa, resulting in estimated parameters being dependent on imputed values. The simultaneous imputation of multiple variables becomes more challenging when missingness is nonmonotone.

Consider the following example. Variable X_1 is complete in observation 1 and missing in observation 2, and variable X_2 is missing in observation 1 and complete in observation 2. We need to impute the two variables simultaneously. Suppose that we impute variable X_2 using previously imputed variable X_1. Observation 1, containing an imputed value of X_1, is used to estimate the model parameters for X_2. As a result, the model parameters (and subsequently imputed values of X_2) are obtained by treating the imputed value of X_1 as if it were true and thus ignore the imputation variability in X_1.

One approach for dealing with an arbitrary missing-value pattern is to assume an explicit tractable parametric model for the data and draw imputed values from the resulting distribution of the missing data given observed data. One of the more popular parametric models is the Gaussian normal model; see Rubin (1987) for other recommendations. Although a multivariate normal model is straightforward, difficulty arises in the simulation from the corresponding, more complicated, distribution of the missing data. One solution is to use one of the Bayesian iterative Markov chain Monte Carlo (MCMC) procedures to approximate the distribution of missing data.

Multivariate imputation using data augmentation

`mi impute mvn` uses data augmentation (DA) —an iterative MCMC procedure—to generate imputed values assuming an underlying multivariate normal model. For details about DA as a general MCMC procedure, see Gelman et al. (2004), Tanner and Wong (1987), and Li (1988), among others. For applications of DA to incomplete multivariate normal data, see, for example, Little and Rubin (2002) and Schafer (1997). Below we briefly describe the idea behind DA; see *Methods and formulas* for details.

Consider multivariate data $\mathbf{X} = (\mathbf{X}_o, \mathbf{X}_m)$, decomposed into the observed part $\mathbf{X}_o$ and the missing part $\mathbf{X}_m$, from a normal distribution $\Pr(\mathbf{X}|\boldsymbol{\theta}) = N(\boldsymbol{\beta}, \boldsymbol{\Sigma})$, where $\boldsymbol{\theta}$ denotes the unknown model parameters (regression coefficients $\boldsymbol{\beta}$ and unique elements of the covariance matrix $\boldsymbol{\Sigma}$). The goal is to replace missing values in $\mathbf{X}_m$ with draws from the distribution (or the predictive distribution in Bayesian terminology) of the missing data given observed data, $\Pr(\mathbf{X}_m|\mathbf{X}_o)$. The actual predictive distribution $\Pr(\mathbf{X}_m|\mathbf{X}_o)$ is difficult to draw from directly because of an underlying dependence on the posterior distribution of the unknown parameters $\boldsymbol{\theta}$, $\Pr(\boldsymbol{\theta}|\mathbf{X}_o)$.

Originally, DA was used to approximate the posterior distribution of the model parameters, $\Pr(\boldsymbol{\theta}|\mathbf{X}_o)$, in Bayesian applications with incomplete data. The idea of DA is to augment the observed data, $\mathbf{X}_o$, with the latent (unobserved) data, $\mathbf{X}_m$, such that the conditional posterior distribution $\Pr(\boldsymbol{\theta}|\mathbf{X}_o, \mathbf{X}_m)$ becomes more tractable and easier to simulate from. Then the procedure becomes as follows. For a current $\boldsymbol{\theta}^{(t)}$, draw $\mathbf{X}_m^{(t+1)}$ from its conditional predictive distribution given the observed data and $\boldsymbol{\theta}$, $\Pr(\mathbf{X}_m|\mathbf{X}_o, \boldsymbol{\theta}^{(t)})$. Next draw $\boldsymbol{\theta}^{(t+1)}$ from its conditional posterior distribution given the augmented data, $\Pr(\boldsymbol{\theta}|\mathbf{X}_o, \mathbf{X}_m^{(t+1)})$. Continue to iterate until the sequence $\{(\mathbf{X}_m^{(t)}, \boldsymbol{\theta}^{(t)}) : t = 1, 2, \ldots\}$, an MCMC sequence, converges to a stationary distribution $\Pr(\boldsymbol{\theta}, \mathbf{X}_m|\mathbf{X}_o)$. This way a complicated task of simulating from $\Pr(\boldsymbol{\theta}|\mathbf{X}_o)$ is replaced by a sequence of simpler simulation tasks of iteratively sampling from $\Pr(\boldsymbol{\theta}|\mathbf{X}_o, \mathbf{X}_m)$ and $\Pr(\mathbf{X}_m|\mathbf{X}_o, \boldsymbol{\theta})$. How is this procedure related to imputation? The sequence $\{\mathbf{X}_m^{(t)} : t = 1, 2, \ldots\}$ contains draws from an approximate predictive distribution $\Pr(\mathbf{X}_m|\mathbf{X}_o)$, and thus $\mathbf{X}_m^{(t)}$'s are, in fact, imputations. The convergence of this procedure was studied by Li (1988).

The functional forms of the conditional distributions $\Pr(\boldsymbol{\theta}|\mathbf{X}_o, \mathbf{X}_m)$ and $\Pr(\mathbf{X}_m|\mathbf{X}_o, \boldsymbol{\theta})$ are determined from the assumed distribution of the data, $\mathbf{X}$, and a prior distribution for the model parameters, $\boldsymbol{\theta}$, $\Pr(\boldsymbol{\theta})$. mi impute mvn assumes a normal distribution for the data and supports three prior distributions: uniform, Jeffreys, and ridge.

The prior distributions are categorized into noninformative (or also vague, diffuse, flat, reference) and informative prior distributions. The noninformative priors provide no extra information about model parameters beyond that already contained in the data. These priors are recommended when no strong prior knowledge is available about the parameters. Informative prior distributions are used when there is some a priori knowledge about the distribution of the parameters. For example, prior information about cancer mortality rates in a Poisson model can be assigned based on the available worldwide estimate. The uniform and Jeffreys priors are noninformative priors. The ridge prior is an informative prior.

The uniform prior assumes that all values of the parameters are equally probable. Under this prior specification, the posterior distribution of the parameters is equivalent to the likelihood function, and so the Bayesian and frequentist methods coincide. The Jeffreys prior is another widely used noninformative prior distribution, and with small samples, it may be preferable to the uniform prior. A ridge prior is often used when the estimated covariance matrix becomes singular (or nearly singular), as may occur with sparse missing data if there are not enough observations to estimate reliably all aspects of the covariance matrix. A ridge prior smooths the estimate of the covariance matrix toward a diagonal structure depending on the chosen degrees of freedom; the larger the degrees of freedom, the closer is the estimated covariance matrix to the diagonal matrix (see Schafer [1997, 155–157] for details).

Convergence of the MCMC method

The MCMC procedure, DA, is iterated until an MCMC sequence $\{(\mathbf{X}_m^{(t)}, \boldsymbol{\theta}^{(t)}) : t = 1, 2, \ldots\}$ converges to a stationary distribution. Unlike maximum likelihood, EM, or other optimization-based procedures, the DA procedure does not have a simple stopping rule that guarantees the convergence of the chain to a stationary distribution. Thus the question of how long to iterate to achieve convergence arises. In addition to determining convergence of MCMC, we must also investigate the serial dependence known to exist among the MCMC draws to obtain independent imputations.

Suppose that after an initial burn-in period, b, the sequence $\{(\mathbf{X}_m^{(b+t)}) : t = 1, 2, \ldots\}$ (imputations) can be regarded as an approximate sample from $\Pr(\mathbf{X}_m|\mathbf{X}_o)$. In general, this sample will not contain independent observations because the successive iterates of the MCMC tend to be correlated. To achieve independence among imputations, we can sample the chain. To do that, we need to determine the number of iterations, k, such that $\mathbf{X}_m^{(t)}$ and $\mathbf{X}_m^{(t+k)}$ are approximately independent. Then imputations can be obtained as the chain values of $\mathbf{X}_m$ from iterations $b, b+k, b+2k, \ldots, b+mk$, where m is the required number of imputations. In our definition, b is the number of iterations necessary for the chain to achieve stationarity and k is the number of iterations between imputations necessary to achieve independent values of the chain.

Before we proceed, we notice that from the properties of MCMC, the convergence of the chain $\{(\mathbf{X}_m^{(t)}, \boldsymbol{\theta}^{(t)}) : t = 1, 2, \ldots\}$ to $\Pr(\boldsymbol{\theta}, \mathbf{X}_m | \mathbf{X}_o)$, is equivalent to the convergence of $\{(\boldsymbol{\theta}^{(t)}) : t = 1, 2, \ldots\}$ to $\Pr(\boldsymbol{\theta} | \mathbf{X}_o)$ or, alternatively, of $\{(\mathbf{X}_m^{(t)}) : t = 1, 2, \ldots\}$ to $\Pr(\mathbf{X}_m | \mathbf{X}_o)$. Because the parameter series are usually of lower dimension, we examine convergence using the series of parameter estimates rather than the series of imputations.

How to determine convergence and, in particular, to choose values for b and k, has received much attention in the MCMC literature. In practice, convergence is often examined visually from the trace and autocorrelation plots of the estimated parameters. Trace plots are plots of estimated parameters against iteration numbers. Long-term trends in trace plots and high serial dependence in autocorrelation plots are indicative of a slow convergence to stationarity. A value of b can be inferred from a trace plot as the earliest iteration after which the chain does not exhibit a visible trend and the parameter series stabilize, which is to say, the fluctuations in values become more regular. A value of k can be chosen from autocorrelation plots as the lag k for which autocorrelations of all parameters decrease to zero. When the initial values are close to the posterior mode, the initial number of iterations, b, and number of iterations between imputations, k, will be similar. When the initial values are far off in the tail of the posterior distribution, the initial number of iterations will generally be larger.

In practice, when the number of parameters in the model is large, it may not be feasible to monitor the convergence of all the individual series. One solution is to find a function of the parameters that would be the slowest to converge to stationarity. The convergence of the series for this function will then be indicative of the convergence of other functions and, in particular, individual parameter series. Schafer (1997, 129–131) suggests the worst linear function (WLF), the function corresponding to the linear combination of the parameter estimates where the coefficients are chosen such that this function has the highest asymptotic rate of missing information; see *Method and formulas* for computational details. He found that when the observed-data posterior distribution is nearly normal, this function is among the slowest to approach stationarity. Thus we can determine b and k by monitoring the convergence of this function. When the observed-data posterior is not normal and some aspects of the model are poorly estimated, this function may not be the slowest to converge. In such cases, we recommend exploring convergence of other functions or of individual parameter series.

The number of iterations necessary for DA to converge depends on the rate of convergence of DA. The rate of convergence of DA mainly depends on the fractions of missing information and initial values. The higher the fractions of missing information and the farther the initial values are from the posterior mode, the slower the convergence, and thus the larger the number of iterations required. Initial values for the DA procedure can be obtained from the EM algorithm for incomplete data (e.g., Dempster, Laird, and Rubin [1977]). In addition, the number of iterations necessary for the DA procedure to converge can be inferred based on the number of iterations that the EM algorithm took to converge (Schafer 1997).

The convergence of the chain and the required number of iterations can be also inferred by running multiple independent MCMC sequences using overdispersed initial values, that is, initial values from a distribution with greater variability than that of the posterior distribution (Gelman and Rubin 1992; Schafer 1997, 126–128). Then the number of iterations can be taken to be the largest iteration number for which the series in all of the chains stabilize.

Although the graphical summaries described above are useful in checking convergence, they must be used with caution. They can be deceptive in cases when the observed-data posterior has an odd shape or has multiple modes, which may happen with small sample sizes or sparse missing data. Examination of the data and missing-data patterns, as well as the behavior of the EM algorithm, are highly recommended when investigating the MCMC convergence. How one checks for convergence will be shown in examples 2 and 4.

Using mi impute mvn

mi impute mvn imputes missing data using DA, an iterative MCMC method, assuming the multivariate normal distribution for the data. For the discussion of options, such as add() and replace, common to all imputation methods, see [MI] **mi impute**. Here we focus on the options and functionality specific to mi impute mvn.

The two main options are burnin() (which specifies the number of iterations necessary for the MCMC to converge, b) and burnbetween() (which specifies the number of iterations between imputations, k). We discussed how to choose these values in the section above. By default, these values are arbitrarily set to be 100 each.

You can choose from the three prior specifications. You can use prior(uniform) (the default) to specify the uniform prior, prior(jeffreys) to specify the Jeffreys prior, or prior(ridge, df()) to specify a ridge prior. You must also choose the degrees of freedom with a ridge prior.

For initial values, mi impute mvn uses the estimates from the EM algorithm for incomplete data (initmcmc(em)). When the uniform prior distribution is used, the estimates obtained from EM are MLEs. Under other prior specifications, the estimates from EM correspond to the posterior mode of the respective posterior distribution of the model parameters. Using the estimates from EM as initial values in general accelerates the convergence of MCMC. To determine convergence, it may also be useful to try different sets of initial values. You can do this by creating Stata matrices containing the initial values and supplying them in the respective initmcmc() suboptions betas(), cov(), etc.

You can save the estimates of the WLF and parameter series from MCMC iterations by using the savewlf() and savetrace() options. These options are useful when examining convergence of MCMC, as we will demonstrate in examples 2 and 4. You can use mi impute mvn to run the MCMC without imputing the data if you specify the mcmconly option. This option is useful in combination with savewlf() or savetrace() when examining convergence of MCMC. When mcmconly is specified, the DA procedure is performed for the number of iterations as specified in burnin() and no imputations are performed.

You can also perform the EM estimation without MCMC iterations if you specify the emonly() option. This option is useful for detecting convergence problems prior to running MCMC. The number of iterations EM takes to converge can be used as an approximation for the burn-in period. Also, slow convergence of the EM algorithm can reveal problems with estimability of certain model parameters.

Examples

▷ Example 1: Monotone-missing data

Recall the heart attack example from *Multivariate imputation* in [MI] **mi impute**, where we used mi impute mvn to impute missing values for age and bmi that follow a monotone-missing pattern:

```
. use http://www.stata-press.com/data/r11/mheart5s0
(Fictional heart attack data; bmi and age missing)
. mi impute mvn age bmi = attack smokes hsgrad female, add(10)
Performing EM optimization:
note: 12 observations omitted from EM estimation because of all imputation
      variables missing
  observed log likelihood = -651.75868 at iteration 7
Performing MCMC data augmentation ...
Multivariate imputation                 Imputations =         10
Multivariate normal regression                added =         10
Imputed: m=1 through m=10                   updated =          0
Prior: uniform                           Iterations =       1000
                                            burn-in =        100
                                            between =        100
```

	Observations per m			
Variable	complete	incomplete	imputed	total
age	142	12	12	154
bmi	126	28	28	154

```
(complete + incomplete = total; imputed is the minimum across m
 of the number of filled in observations.)
```

Compared with the example in [MI] **mi impute**, we omit the nolog option in the above.

In addition to the output reported by all imputation methods, mi impute mvn also provides some specific information.

As we previously explained, mi impute mvn uses an iterative MCMC technique to impute missing values. The two phases of mi impute mvn are 1) obtaining initial values (unless supplied directly) and 2) performing the MCMC procedure from which imputations are obtained. These two phases are noted in the output header.

In the above example, the initial values are obtained using the EM method (the default). We see from the output that EM converged in seven iterations. A note displayed thereafter reports that 12 observations contain missing values for both bmi and age and were omitted. The note is just explanatory and should not cause you concern. Those 12 observations would contribute nothing to the likelihood function even if they were included, although the algorithm would take longer to converge.

The estimates from EM are used as initial values for DA. The first part of the table header, containing the information about the method used and the number of imputations, was described in detail in [MI] **mi impute**. The second part of the table header is specific to mi impute mvn. From the output, a total of 1,000 iterations of MCMC are performed. The first 100 iterations (the default) are used for the burn-in period (burn-in = 100), the first imputation calculated from the last iteration; thereafter, each subsequent imputation is calculated after performing another 100 iterations. The default uniform prior is used for both the EM estimation and the MCMC procedure. Under this prior, the parameter estimates obtained are MLEs.

◁

▷ Example 2: Checking convergence of MCMC

In the above example, the monotone missingness of age and bmi as well as the quick convergence of EM suggest that the MCMC must converge rapidly. In fact, we know that under a monotone-missing pattern, no iterations are needed to obtain imputed values (see [MI] **mi impute monotone**). Let's examine the convergence of the MCMC procedure for the above heart attack data, the point being to see what quick convergence looks like.

As we discussed earlier, convergence is often assessed from the trace plots of the MCMC parameter estimates. Because of a possibly large number of estimated parameters, this approach may be tedious. Alternatively, we can plot the WLF for which the convergence is generally the slowest.

We use the `savewlf(wlf)` option to save estimates of the WLF to a Stata dataset called `wlf.dta`. To examine the convergence of MCMC, we do not need imputation, and so we use the `mcmconly` option to perform the MCMC procedure without subsequent imputation. We use a total of $1000 = 10 \times 100$ iterations (`burnin(1000)` option), corresponding to the length of the MCMC to obtain 10 imputations:

```
. mi impute mvn age bmi = attack smokes hsgrad female, mcmconly burnin(1000)
> rseed(2232) savewlf(wlf)

Performing EM optimization:
note: 12 observations omitted from EM estimation because of all imputation
      variables missing
   observed log likelihood = -651.75868 at iteration 7

Performing MCMC data augmentation ...

Note: no imputation performed.
```

We also specified the `rseed(2232)` option so that we can reproduce our results.

The created dataset contains three variables: `iter`, `m`, and `wlf`. The `iter` variable records iterations (the burn-in iterations are recorded as negative integers). The `m` variable records imputation numbers to which the iteration sequence corresponds (m contains 0 if mcmconly is used). The `wlf` variable records the WLF estimates.

```
. use wlf, clear

. describe

Contains data from wlf.dta
  obs:         1,000
 vars:             3                          24 Jun 2009 21:28
 size:        20,000 (99.8% of memory free)
----------------------------------------------------------------
              storage  display    value
variable name   type   format     label      variable label
----------------------------------------------------------------
iter            long   %12.0g
m               long   %12.0g
wlf             double %10.0g
----------------------------------------------------------------
Sorted by:
```

We use the time-series commands `tsline` and `ac` (see [TS] **tsline** and [TS] **corrgram**) to plot the estimates and autocorrelations of `wlf` with respect to the iteration number. We first use `tsset` to set `iter` as the "time" variable and then use `tsline` to obtain a trace plot:

```
. tsset iter
        time variable:  iter, -999 to 0
                delta:  1 unit
. tsline wlf, ytitle(Worst linear function) xtitle(Burn-in period)
```

[Graph: Worst linear function vs Burn-in period from -1000 to 0]

The graph shows no visible trend in the estimates of the WLF, just as we expected. Convergence of MCMC by the 100th iteration should be assured. In fact, taking into account the declared convergence of the EM algorithm in only seven iterations, we would be comfortable with using a much smaller burn-in period of, say, 10 iterations.

We next examine the autocorrelation in the WLF to obtain an idea of how many iterations to use between imputations to ensure their approximate independence:

```
. ac wlf, title(Worst linear function) ytitle(Autocorrelations)
> ciopts(astyle(none)) note("")
```

[Graph: Autocorrelations of Worst linear function vs Lag from 0 to 40]

From the graphical output, the autocorrelations die off quickly (notice a rapid decrease after the first-lag autocorrelation). This suggests that we can use a smaller number, say, 10 or 20, rather than the default 100 iterations for the burn-between period.

◁

▷ Example 3: Arbitrary missing-data pattern

We considered an example with a monotone-missing pattern. `mi impute mvn` is designed to accommodate arbitrary missing-data patterns.

Consider data on house resale prices provided by the Albuquerque Board of Realtors and distributed by the Data and Story Library. You can find a detailed description of the data at http://lib.stat.cmu.edu/DASL/Stories/homeprice.html.

```
. use http://www.stata-press.com/data/r11/mhouses1993
(Albuquerque Home Prices Feb15-Apr30, 1993)

. describe

Contains data from http://www.stata-press.com/data/r11/mhouses1993.dta
  obs:           117                        Albuquerque Home Prices
                                              Feb15-Apr30, 1993
 vars:             8                        19 Jun 2009 10:50
 size:         1,755 (99.9% of memory free) (_dta has notes)

              storage  display    value
variable name   type   format     label      variable label

price           int    %8.0g                 Sale price (hundreds)
sqft            int    %8.0g                 Square footage of living space
age             byte   %10.0g                Home age (years)
nfeatures       byte   %8.0g                 Number of certain features
ne              byte   %8.0g                 Located in northeast (largest
                                               residential) sector of the city
custom          byte   %8.0g                 Custom build
corner          byte   %8.0g                 Corner location
tax             int    %10.0g                Tax amount (dollars)

Sorted by:
```

The dataset includes eight variables. The primary variable of interest is `price`, and other variables are used as its predictors.

We investigate the missing-data patterns of these data using `misstable`:

```
. misstable pattern

  Missing-value patterns
   (1 means complete)

               |  Pattern
      Percent  |  1  2

         56%   |  1  1

          35   |  1  0
           7   |  0  0
           2   |  0  1

        100%   |

Variables are  (1) tax  (2) age

. misstable nested
    1.  tax(10)
    2.  age(49)
```

We see from the output only 56% of observations are complete; the remaining 44% contain missing values of `age` or `tax`. The `tax` variable contains 10 missing values, and the `age` variable contains 49 missing values. `misstable nested` reports that missing values of `age` and `tax` are not nested because there are two statements describing the missing-value pattern; see [R] **misstable** for details.

Let's use `mi impute mvn` to impute missing values of `age` and `tax`. Before we do that, a quick examination of the data revealed that the distribution for `age` and `tax` are somewhat skewed. As such, we choose to impute the variables on a log-transformed scale.

Following the steps as described in *Imputing transformations of imputed variables* of [MI] **mi impute**, we create new variables containing the log values,

```
. gen lnage = ln(age)
(49 missing values generated)
. gen lntax = ln(tax)
(10 missing values generated)
```

and register them as imputed variables,

```
. mi set mlong
. mi register imputed lnage lntax
(51 m=0 obs. now marked as incomplete)
. mi register regular price sqft nfeatures ne custom corner
```

We `mi set` our data as mlong and register the complete variables as regular. For the purpose of this analysis, we leave passive variables `age` and `tax` unregistered. (Note that all missing values of the created `lnage` and `lntax` variables are eligible for imputation; see [MI] **mi impute** for details.)

We now use `mi impute mvn` to impute values of `lnage` and `lntax`:

```
. mi impute mvn lnage lntax = price sqft nfeatures ne custom corner, add(20)
Performing EM optimization:
note: 8 observations omitted from EM estimation because of all imputation
      variables missing
  observed log likelihood =   112.1464 at iteration 48
Performing MCMC data augmentation ...
```

Multivariate imputation		Imputations =	20
Multivariate normal regression		added =	20
Imputed: m=1 through m=20		updated =	0
Prior: uniform		Iterations =	2000
		burn-in =	100
		between =	100

	Observations per m			
Variable	complete	incomplete	imputed	total
lnage	68	49	49	117
lntax	107	10	10	117

(complete + incomplete = total; imputed is the minimum across m
of the number of filled in observations.)

◁

▷ Example 4: Checking convergence of MCMC

In the above example, we arbitrarily created 20 imputations. The output is similar to that of the earlier example. Here the EM algorithm converges by the 48th iteration. This suggests that, again, the default 100 iterations for the burn-in period should be sufficient for the convergence of MCMC. Nevertheless, we choose to confirm this visually by repeating the steps from example 2.

We run the MCMC for a total of 2,000 iterations (as would be necessary to obtain 20 imputations) without imputing data and set the seed for reproducibility. We overwrite the existing `wlf.dta` file to contain the new estimates of the WLF by specifying `replace` within `savewlf()`:

```
. mi impute mvn lnage lntax = price sqft nfeatures ne custom corner,
> mcmconly burnin(2000) rseed(23) savewlf(wlf, replace)
Performing EM optimization:
note: 8 observations omitted from EM estimation because of all imputation
      variables missing
  observed log likelihood =    112.1464 at iteration 48
Performing MCMC data augmentation ...
Note: no imputation performed.
```

We generate the same graphs as in example 2, this time using the new estimates of the WLF:

```
. use wlf, clear
. tsset iter
        time variable:  iter, -1999 to 0
                delta:  1 unit
. tsline wlf, ytitle(Worst linear function) xtitle(Burn-in period)
```

```
. ac wlf, title(Worst linear function) ytitle(Autocorrelations)
> ciopts(astyle(none)) note("")
```

Compared with the earlier graphs, the time-series graphs do not reveal any apparent trend, but the autocorrelation dies out more slowly. The default values of 100 for the initial burn-in and between-imputation iterations should be sufficient.

◁

▷ Example 5: Alternative prior distribution

Consider hypothetical data:

```
. use http://www.stata-press.com/data/r11/mvnexample0, clear
(Fictional data for -mi impute mvn-)

. mi describe
  Style:  mlong
          last mi update 19jun2009 10:50:22, 2 days ago
  Obs.:   complete             3
          incomplete          17   (M = 0 imputations)
          ─────────────────────
          total               20
  Vars.:  imputed:  3; x1(16) x2(5) x3(17)
          passive:  0
          regular:  0
          system:   1; _mi_miss
          (there are no unregistered variables)
```

Continuous normally-distributed variables x1, x2, and x3 contain missing values. For illustration purposes, we consider an extreme case when some variables (x1 and x3 here) contain only a few complete observations.

We use mi impute mvn to impute missing values and create 30 imputations. Notice that in this example, we do not have complete predictors, and so the right-hand-side specification is empty:

```
. mi imp mvn x1-x3, add(30) rseed(332242)
Performing EM optimization:
note: 4 observations omitted from EM estimation because of all imputation
      variables missing
  observed log likelihood = 6.5368927 at iteration 100
  (EM did not converge)
Performing MCMC data augmentation ...
Iteration 94: variance-covariance matrix (Sigma) became not positive definite
posterior distribution is not proper
r(498);
```

mi impute mvn terminates with an error reporting that the estimated variance–covariance matrix became nonpositive definite. mi impute mvn terminated because the posterior predictive distribution of missing data is not proper, but notice also that EM did not converge after the default 100 iterations.

There are two issues here. First, because EM did not converge after 100 iterations, we suspect that the default 100 iterations used for the burn-in period may not be large enough for MCMC to converge. Second, the observed missing-data pattern presents difficulties with estimating the covariance matrix reliably, which leads to a nonpositive-definite estimate during the MCMC iteration.

The first issue may be resolved by increasing the maximum number of iterations for EM by using EM's iterate() suboption. Convergence of EM, however, does not guarantee convergence of the MCMC by the same number of iterations. For one, the convergence of EM is relative to the specified tolerance, and more stringent conditions may lead to a nonconvergent result. As such, we recommend that you always examine the obtained MCMC results.

The second issue is not surprising. Recall that x1 and x3 have very few complete observations. So the aspects of the covariance structure involving those variables (for example, the covariance between x1 and x2) are difficult to estimate reliably based on the information from the observed data only. The default uniform prior may not be viable here.

One solution is to introduce prior information to stabilize the estimation of the covariance matrix. We can do this by specifying a ridge prior using the `prior()` option. We introduce only a small amount of information by using a degrees of freedom value of 0.1:

```
. mi imp mvn x1-x3, add(30) prior(ridge, df(0.1)) rseed(332242)

Performing EM optimization:
note: 4 observations omitted from EM estimation because of all imputation
      variables missing
  observed log posterior =    -1.13422 at iteration 100
  (EM did not converge)

Performing MCMC data augmentation ...

Multivariate imputation              Imputations  =        30
Multivariate normal regression             added  =        30
Imputed: m=1 through m=30                updated  =         0

Prior: ridge, df=.1                   Iterations  =      3000
                                         burn-in  =       100
                                         between  =       100

                      |          Observations per m
                      |---------------------------------------
             Variable | complete   incomplete   imputed | total
             ---------|------------------------------------------
                   x1 |        4           16        16 |    20
                   x2 |       15            5         5 |    20
                   x3 |        3           17        17 |    20

(complete + incomplete = total; imputed is the minimum across m
 of the number of filled in observations.)
```

This appears to be enough to alleviate the problem of a nonpositive-definite estimate of the covariance matrix. Still, EM did not converge.

We will fix that and examine the resulting MCMC sequence. We will use the same random-number seed and this time save the WLF. Rather than imputing the data as before, we will simply run the MCMC for the same number of iterations it takes to obtain 30 imputations using the default settings, namely, $30 \times 100 = 3000$.

```
. mi imp mvn x1-x3, mcmconly prior(ridge, df(0.1)) initmcmc(em, iter(200) nolog)
> burnin(3000) savewlf(wlf, replace) rseed(332242)

Performing EM optimization:
note: 4 observations omitted from EM estimation because of all imputation variables
      missing
  observed log posterior = -1.1341806 at iteration 152

Performing MCMC data augmentation ...

Note: no imputation performed.
```

We increased the maximum number of iterations for the EM algorithm to 200; it converged in iteration 152.

We use the results from `wlf.dta` to obtain the trace and autocorrelation plots as we did in the earlier examples:

The serial correlation decreases slowly. There is no obvious trend in the WLF estimates, but we notice high variability and several spikes, some distinctive. The high variability and spikes are not surprising considering that certain model parameters could not be estimated reliably from the observed data and considering that we did not introduce enough prior information to obtain less variable estimates; we introduced only enough to achieve nonsingularity.

We could decrease the variability of the estimates by obtaining more data or introducing stronger prior information. For example, we could increase the number of degrees of freedom with a ridge prior to constrain the covariance matrix toward a diagonal structure:

```
. mi imp mvn x1-x3, replace prior(ridge, df(10)) burnin(300) rseed(332242)
  (output omitted)
```

If we create and examine the trace plots and autocorrelations of the WLF under the new prior specification, we find that variability of the estimates and serial dependence decrease greatly at a cost of bias if the prior assumptions are false.

◁

▷ Example 6: Saving all parameter series

The examples above used the WLF to monitor convergence of MCMC because in most applications it is sufficient. Although the WLF series often behave as the worst-case scenario, exceptions to this exist in practice. Sometimes, examining individual parameter series may be necessary.

We can save all parameter series from MCMC by using the `saveptrace()` option. These parameter series are saved in a parameter-trace file, a special file with extension .stptrace. Although the resulting file is not a Stata dataset, it can be easily loaded into Stata using `mi ptrace use`; see [MI] **mi ptrace** for details.

Let's look at several parameter series from the above example.

```
. use mvn
. mi impute mvn x1-x3, mcmconly prior(ridge, df(0.1)) init(em, iter(200) nolog)
> burnin(3000) rseed(332242) saveptrace(parms)
(output omitted)
```

We save all parameter series to a file called `parms` by using `stptrace(parms)`.

We first describe the contents of the `parms` file and then read it into Stata:

```
. mi ptrace describe parms
file parms.stptrace created on 24 Jun 2009 21:28 contains 3,000 records
(obs.) on
    m                     1 variable
    iter                  1 variable
    b[y, x]               3 variables (3 x 1)
    v[y, y]               6 variables (3 x 3, symmetric)
  where y and x are
    y:  (1) x1   (2) x2   (3) x3
    x:  (1) _cons

. mi ptrace use parms, clear
```

The output from `mi ptrace describe` reports that the file contains imputation numbers, iteration numbers, estimates of three regression coefficients (b[x1,_cons], b[x2,_cons], b[x3,_cons], which are effectively the means of x1, x2, and x3), and estimates of six covariances (v[x1,x1], v[x2,x1], v[x2,x2], and so on).

Because x1 and x3 contain the least number of complete observations, we examine the series containing their variance and covariance estimates.

```
. tsset iter
        time variable:  iter, -2999 to 0
                delta:  1 unit
. tsline v_y1y1, name(gr1) nodraw ytitle(Var(x1)) xtitle("") ylabel(#4)
. tsline v_y3y1, name(gr2) nodraw ytitle(Cov(x3,x1)) xtitle("") ylabel(#4)
. tsline v_y3y3, name(gr3) nodraw ytitle(Var(x3)) xtitle("") ylabel(#4)
. graph combine gr1 gr2 gr3, xcommon cols(1) b1title(Iteration)
```

We generate graphs separately for each series and then combine them in one graph using graph combine; see [G] **graph combine**.

We repeat the same for the autocorrelation graphs:

```
. ac v_y1y1, ytitle(Var(x1)) xtitle("") ciopts(astyle(none)) note("")
> name(gr1, replace) nodraw ylabel(#4)
. ac v_y3y1, ytitle(Cov(x3,x1)) xtitle("") ciopts(astyle(none)) note("")
> name(gr2, replace) nodraw ylabel(#4)
. ac v_y3y3, ytitle(Var(x3)) xtitle("") ciopts(astyle(none)) note("")
> name(gr3, replace) nodraw ylabel(#4)
. graph combine gr1 gr2 gr3, xcommon cols(1) title(Autocorrelations) b1title(Lag)
```

We can see that the trace plot and autocorrelations corresponding to the variance of x1 resemble the patterns of the earlier WLF estimates. We also notice that the series for the variance of x3 have slightly higher serial dependence than the other two.

Again, if we switch to using a ridge prior with 10 degrees of freedom and repeat the steps above, the obtained trace plots will be more precise and more regular. The serial dependence in the series will be lower.

◁

Saved results

mi impute mvn saves the following in r():

Scalars
r(M)	total number of imputations
r(M_add)	number of added imputations
r(M_update)	number of updated imputations
r(k_ivars)	number of imputed variables
r(burnin)	number of burn-in iterations
r(burnbetween)	number of burn-between iterations
r(df_prior)	prior degrees of freedom (saved only with prior(ridge))
r(N_em)	number of observations used by EM (without omitted missing observations)
r(N_e_em)	number of observations used by EM in estimation (with omitted missing observations)
r(N_mis_em)	number of incomplete observations within the EM estimation sample
r(N_S_em)	number of unique missing-value patterns
r(niter_em)	number of iterations EM takes to converge
r(llobs_em)	observed log likelihood (saved with prior(uniform))
r(lpobs_em)	observed log posterior (saved with priors other than uniform)
r(converged_em)	convergence flag for EM
r(emonly)	1 if performed EM estimation only, 0 otherwise
r(mcmconly)	1 if performed MCMC only without imputing data, 0 otherwise

Macros
r(method)	name of imputation method (mvn)
r(ivars)	names of imputation variables
r(rseed)	random-number seed
r(prior)	prior distribution
r(init_mcmc)	type of initial values (em or user)
r(ivarsorder)	names of imputation variables in the order used in the computation
r(init_em)	type of initial values used by EM (ac, cc, or user)

Matrices
r(N)	number of observations in imputation sample (per variable)
r(N_complete)	number of complete observations in imputation sample (per variable)
r(N_incomplete)	number of incomplete observations in imputation sample (per variable)
r(N_imputed)	number of imputed observations in imputation sample (per variable)
r(Beta0)	initial values for regression coefficients used by DA
r(Sigma0)	initial variance–covariance matrix used by DA
r(wlf_wgt)	coefficients for the WLF (saved with initmcmc(em) or if wlfwgt() is used)
r(Beta_em)	estimated regression coefficients from EM
r(Sigma_em)	estimated variance–covariance matrix from EM
r(Beta0_em)	initial values for regression coefficients used by EM
r(Sigma0_em)	initial variance–covariance matrix used by EM
r(N_pat)	minimum, average, and maximum numbers of observations per missing-value pattern

r(N_pat) and results with the _em suffix are saved only when the EM algorithm is used (with emonly or initmcmc(em)).

Methods and formulas

Let $\mathbf{x}_1, \mathbf{x}_2, \ldots, \mathbf{x}_N$ be a random sample from a p-variate normal distribution recording values of p imputation variables. Consider a multivariate normal regression

$$\mathbf{x}_i = \boldsymbol{\Theta}' \mathbf{z}_i + \boldsymbol{\epsilon}_i, \; i = 1, \ldots, N$$

where $\mathbf{z}_i$ is a $q \times 1$ vector of independent (complete) variables from observation i, $\boldsymbol{\Theta}$ is a $q \times p$ matrix of regression coefficients, and $\boldsymbol{\epsilon}_i$ is a $p \times 1$ vector of random errors from a p-variate normal distribution with a zero mean vector and a $p \times p$ positive-definite covariance matrix $\boldsymbol{\Sigma}$. We refer to $\boldsymbol{\Theta}$ and $\boldsymbol{\Sigma}$ as model parameters. Consider the partition $\mathbf{x}_i = (\mathbf{x}_{i(\mathrm{m})}, \mathbf{x}_{i(\mathrm{o})})$ corresponding to missing and observed values of imputation variables in observation i for $i = 1, \ldots, N$.

Methods and formulas are presented under the following headings:

> Data augmentation
> Prior distribution
> Initial values: EM algorithm
> Worst linear function

Data augmentation

`mi impute mvn` uses data augmentation (DA) to fill in missing values in $\mathbf{x}_i$ independently for each observation $i = 1, \ldots, N$. Data augmentation consists of the two steps, an I step (imputation step) and a P step (posterior step), performed at each iteration $t = 0, 1, \ldots, T$. At iteration t of the I step, missing values in $\mathbf{x}_i$ are replaced with draws from the conditional posterior distribution of $\mathbf{x}_{i(\mathrm{m})}$ given observed data and current values of model parameters independently for each $i = 1, \ldots, N$. During the P step, new values of model parameters are drawn from their conditional posterior distribution given the observed data and the data imputed in the previous I step. Mathematically, this process can be described as follows:

I step:

$$\mathbf{x}_{i(\mathrm{m})}^{(t+1)} \sim P\left(\mathbf{x}_{i(\mathrm{m})} | \mathbf{z}_i, \mathbf{x}_{i(\mathrm{o})}, \boldsymbol{\Theta}^{(t)}, \boldsymbol{\Sigma}^{(t)}\right), \; i = 1, \ldots, N \quad (1)$$

P step:

$$\begin{aligned} \boldsymbol{\Sigma}^{(t+1)} &\sim P\left(\boldsymbol{\Sigma} | \mathbf{z}_i, \mathbf{x}_{i(\mathrm{o})}, \mathbf{x}_{i(\mathrm{m})}^{(t+1)}\right) \\ \boldsymbol{\Theta}^{(t+1)} &\sim P\left(\boldsymbol{\Theta} | \mathbf{z}_i, \mathbf{x}_{i(\mathrm{o})}, \mathbf{x}_{i(\mathrm{m})}^{(t+1)}, \boldsymbol{\Sigma}^{(t+1)}\right) \end{aligned} \quad (2)$$

The above two steps are repeated until the specified number of iterations, T, is reached. The total number of iterations, T, is determined by the length of the initial burn-in period, b, and the number of iterations between imputations k. Specifically, $T = b + M_{new} \times k$, where M_{new} contains the number of added and updated imputations. `mi impute mvn` saves imputed values $\mathbf{x}_{i(\mathrm{m})}^{(t_1)}, \mathbf{x}_{i(\mathrm{m})}^{(t_2)}, \ldots, \mathbf{x}_{i(\mathrm{m})}^{(t_{M_{new}})}$ as final imputations, where iteration $t_i = b + (i-1)k$.

By default, `mi impute mvn` uses default values of 100 for b and k. These values may be adequate in some applications and may be too low in others. In general, b and k must be determined based on the properties of the observed Markov chain $\left(\mathbf{X}_{\mathrm{m}}^{(1)}, \boldsymbol{\Theta}^{(1)}, \boldsymbol{\Sigma}^{(1)}\right), \left(\mathbf{X}_{\mathrm{m}}^{(2)}, \boldsymbol{\Theta}^{(2)}, \boldsymbol{\Sigma}^{(2)}\right), \ldots,$ where $\mathbf{X}_{\mathrm{m}}^{(t)}$ denotes all values imputed at iteration t. b must be large enough so that the above chain converges to the stationary distribution $P(\mathbf{X}_{\mathrm{m}}, \boldsymbol{\Theta}, \boldsymbol{\Sigma} | \mathbf{Z}, \mathbf{X}_{\mathrm{o}})$ by iteration $t = b$. k must be large enough so that random draws (imputations) $\mathbf{x}_{i(\mathrm{m})}^{(t_1)}, \mathbf{x}_{i(\mathrm{m})}^{(t_2)}, \ldots$ are approximately independent. See *Convergence of the MCMC method* for more details.

The functional form of the conditional posterior distributions (1)–(2) depends on the distribution of the data and a prior distribution of the model parameters. `mi impute mvn` assumes an improper uniform prior distribution for Θ and an inverted Wishart distribution (Mardia, Kent, and Bibby 1979, 85) $W_p^{-1}(\Lambda, \lambda)$ for Σ under which the prior joint density function is

$$f(\Theta, \Sigma) \propto |\Sigma|^{-(\frac{\lambda+p+1}{2})} \exp\left(-\frac{1}{2}\mathrm{tr}\Lambda^{-1}\Sigma^{-1}\right)$$

Under the multivariate normal model and the above prior distribution, I and P steps become (Schafer 2008; Schafer 1997, 181–185):

I step: $\quad \mathbf{x}_{i(\mathrm{m})}^{(t+1)} \sim N_{p_i}\left(\boldsymbol{\mu}_{\mathrm{m\cdot o}}^{(t)}, \boldsymbol{\Sigma}_{\mathrm{mm\cdot o}}^{(t)}\right), \; i = 1, \ldots, N$

P step: $\quad \boldsymbol{\Sigma}^{(t+1)} \sim W^{-1}(\Lambda_\star^{(t+1)}, \lambda_\star)$

$\quad\quad\quad\quad \mathrm{vec}\left(\Theta^{(t+1)}\right) \sim N_{pq}\left\{\mathrm{vec}\left(\widehat{\Theta}^{(t+1)}\right), \boldsymbol{\Sigma}^{(t+1)} \otimes (\mathbf{Z}'\mathbf{Z})^{-1}\right\}$

where p_i is the number of imputation variables containing missing values in observation i and $\otimes$ is the Kronecker product. Submatrices $\boldsymbol{\mu}_{\mathrm{m\cdot o}}^{(t)}$ and $\boldsymbol{\Sigma}_{\mathrm{mm\cdot o}}^{(t)}$ are the mean and variance of the conditional normal distribution of $\mathbf{x}_{i(\mathrm{m})}$ given $\mathbf{x}_{i(\mathrm{o})}$ based on $(\mathbf{x}_{i(\mathrm{m})}, \mathbf{x}_{i(\mathrm{o})}|\mathbf{z}_i) \sim N_p\left(\Theta^{(t)\prime}\mathbf{z}_i, \boldsymbol{\Sigma}^{(t)}\right)$. See, for example, Mardia, Kent, and Bibby (1979, 63) for the corresponding formulas of the conditional mean and variance of the multivariate normal distribution. The matrix $\widehat{\Theta}^{(t+1)} = (\mathbf{Z}'\mathbf{Z})^{-1}\mathbf{Z}'\mathbf{X}^{(t+1)}$ is the OLS estimate of the regression coefficients based on the augmented data $\mathbf{X}^{(t+1)} = (\mathbf{X}_\mathrm{o}, \mathbf{X}_\mathrm{m}^{(t+1)})$ from iteration t. The posterior cross-product matrix $\Lambda_\star^{(t+1)}$ and the posterior degrees of freedom $\lambda_\star$ are defined as follows:

$$\Lambda_\star^{(t+1)} = \left\{\Lambda^{-1} + (\mathbf{X}^{(t+1)} - \mathbf{Z}\widehat{\Theta}^{(t+1)})'(\mathbf{X}^{(t+1)} - \mathbf{Z}\widehat{\Theta}^{(t+1)})\right\}^{-1}$$

and

$$\lambda_\star = \lambda + N - q$$

Prior distribution

As we already mentioned, `mi impute mvn` assumes an improper uniform prior distribution for Θ and an inverted Wishart distribution for Σ under which the prior joint density function is

$$f(\Theta, \Sigma) \propto |\Sigma|^{-(\frac{\lambda+p+1}{2})} \exp\left(-\frac{1}{2}\mathrm{tr}\Lambda^{-1}\Sigma^{-1}\right)$$

Parameters of the inverted Wishart prior distribution, the prior cross-product matrix Λ, and the prior degrees of freedom λ, are determined based on the requested prior distribution.

By default, `mi impute mvn` uses the uniform prior distribution under which $\lambda = -(p+1)$ and $\Lambda^{-1} = \mathbf{0}_{p\times p}$. Under the uniform prior, the log-likelihood and log-posterior functions are equivalent, and so the ML estimates of the parameters are equal to the posterior mode.

Under the noninformative Jeffreys prior distribution, $\lambda = 0$ and $\Lambda^{-1} = \mathbf{0}_{p\times p}$.

Under a ridge prior distribution, λ is equal to the user-specified value, and $\Lambda^{-1} = \lambda \Sigma_\star$, where the diagonal matrix $\Sigma_\star$ contains the diagonal elements of the estimate of the covariance matrix using all available cases. The variances (diagonal estimates) are the estimates of the mean squared error from regression of each imputation variable on the complete predictors. See Schafer (1997, 155–157) for details. With $\lambda = 0$, this prior specification reduces to the Jeffreys prior.

Initial values: EM algorithm

Initial values $\Theta^{(0)}$ and $\Sigma^{(0)}$ for DA are obtained from the EM algorithm for the incomplete multivariate normal data (for example, Dempster, Laird, and Rubin [1977], Little and Rubin [2002], Schafer [1997]). The EM algorithm iterates between the expectation step (E step) and maximization step (M step) to maximize the log-likelihood (or log-posterior) function.

The observed-data log likelihood is

$$l_l(\Theta, \Sigma | \mathbf{X}_o) = \sum_{s=1}^{S} \sum_{i \in I(s)} \left\{ -0.5 \ln(|\Sigma_s|) - 0.5 (\mathbf{x}_{i(o)} - \Theta'_s \mathbf{z}_i)' \Sigma_s^{-1} (\mathbf{x}_{i(o)} - \Theta'_s \mathbf{z}_i) \right\}$$

where S is the number of unique missing-value patterns, $I(s)$ is the set of observations from the same missing-value pattern s, and Θ_s and Σ_s are the submatrices of Θ and Σ that correspond to the imputation variables, which are observed in pattern s.

The observed-data log posterior is

$$l_p(\Theta, \Sigma | \mathbf{X}_o) = l_l(\Theta, \Sigma | \mathbf{X}_o) + \ln\{f(\Theta, \Sigma)\} = l_l(\Theta, \Sigma | \mathbf{X}_o) - \frac{\lambda + p + 1}{2} \ln(|\Sigma|) - \mathrm{tr}(\Lambda^{-1} \Sigma^{-1})$$

The E step and M step of the EM algorithm are defined as follows (see Schafer [2008; 1997, 163–175] for details).

Let $T_1 = \sum_{i=1}^{N} \mathbf{z}_i \mathbf{x}'_i$ and $T_2 = \sum_{i=1}^{N} \mathbf{x}_i \mathbf{x}'_i$ denote the sufficient statistics for the multivariate normal model. Consider the submatrices $\Theta_{i(o)}$ and $\Theta_{i(m)}$ of Θ, and the submatrices $\Sigma_{i(mm)}$, $\Sigma_{i(mo)}$, and $\Sigma_{i(oo)}$ of Σ corresponding to the observed and missing columns of $\mathbf{x}_i$. Let $O(s)$ and $M(s)$ correspond to the column indexes of the observed and missing parts of $\mathbf{x}_i$ for each missing-values pattern s.

During the E step, the expectations $E(T_1)$ and $E(T_2)$ are computed with respect to the conditional distribution $\Pr(\mathbf{X}_m | \mathbf{X}_o, \Theta^{(t)}, \Sigma^{(t)})$ using the following relations:

$$E(x_{ij} | \mathbf{X}_o, \Theta^{(t)}, \Sigma^{(t)}) = \begin{cases} x_{ij}, & \text{for } j \in O(s) \\ x^\star_{ij}, & \text{for } j \in M(s) \end{cases}$$

and

$$E(x_{ij} x_{il} | \mathbf{X}_o, \Theta^{(t)}, \Sigma^{(t)}) = \begin{cases} x_{ij} x_{il}, & \text{for } j, l \in O(s) \\ x^\star_{ij} x_{il}, & \text{for } j \in M(s), l \in O(s) \\ c_{ij} + x^\star_{ij} x^\star_{il}, & \text{for } j, l \in M(s) \end{cases}$$

where $x^\star_{ij}$ is the jth element of the vector $\Theta'_{i(m)} \mathbf{z}_i + \Sigma_{i(mo)} \Sigma_{i(oo)}^{-1} \left(\mathbf{x}_{i(o)} - \Theta'_{i(o)} \mathbf{z}_i \right)$, and c_{ij} is the element of the matrix $\Sigma_{i(mm)} - \Sigma_{i(mo)} \Sigma_{i(oo)}^{-1} \Sigma'_{i(mo)}$.

During the P step, the model parameters are updated using the computed expectations of the sufficient statistics:

$$\Theta^{(t+1)} = (\mathbf{Z}'\mathbf{Z})^{-1} E(T_1)$$

$$\Sigma^{(t+1)} = \frac{1}{N + \lambda + p + 1} \left\{ E(T_2) - E(T_1)'(\mathbf{Z}'\mathbf{Z})^{-1} E(T_1) + \Lambda^{-1} \right\}$$

EM iterates between the E step and the P step until the maximum relative difference between the two successive values of all parameters is less than the default tolerance of 1e–5 (or the specified `tolerance()`).

Worst linear function

The worst linear function is defined as follows (Schafer 1997, 129–131):

$$\xi(\theta) = \widehat{v}_1'(\theta - \widehat{\theta})$$

where θ and $\widehat{\theta}$ are column vectors of the unique model parameters and their respective EM estimates; $\widehat{v}_1 = \theta^{(t)} - \theta^{(t-1)}$, where $\theta^{(t)} = \widehat{\theta}$ and $\theta^{(t-1)}$ are the estimates from the last and one before the last iterations of the EM algorithm. This function is regarded to be the worst linear function because it has the highest asymptotic rate of missing information among all linear functions. This function is derived based on the convergence properties of the EM algorithm (see Schafer [1997, 55–59] for details).

References

Dempster, A. P., N. M. Laird, and D. B. Rubin. 1977. Maximum likelihood from incomplete data via the EM algorithm. *Journal of the Royal Statistical Society, Series B* 39: 1–38.

Gelman, A., J. B. Carlin, H. S. Stern, and D. B. Rubin. 2004. *Bayesian Data Analysis*. 2nd ed. London: Chapman & Hall/CRC.

Gelman, A., and D. B. Rubin. 1992. Inference from iterative simulation using multiple sequences. *Statistical Science* 7: 457–472.

Li, K.-H. 1988. Imputation using Markov chains. *Journal of Statistical Computation and Simulation* 30: 57–79.

Little, R. J. A., and D. B. Rubin. 2002. *Statistical Analysis with Missing Data*. 2nd ed. Hoboken, NJ: Wiley.

Mardia, K. V., J. T. Kent, and J. M. Bibby. 1979. *Multivariate Analysis*. London: Academic Press.

Rubin, D. B. 1987. *Multiple Imputation for Nonresponse in Surveys*. New York: Wiley.

Schafer, J. L. 1997. *Analysis of Incomplete Multivariate Data*. Boca Raton, FL: Chapman & Hall/CRC.

———. 2008. *NORM: Analysis of incomplete multivariate data under a normal model, Version 3*. Software package for R. University Park, PA: The Methodology Center, Pennsylvania State University.

Tanner, M. A., and W. H. Wong. 1987. The calculation of posterior distributions by data augmentation (with discussion). *Journal of the American Statistical Association* 82: 528–550.

Also see

[MI] **mi impute** — Impute missing values

[MI] **mi impute monotone** — Impute missing values in monotone data

[MI] **mi estimate** — Estimation using multiple imputations

[MI] **intro substantive** — Introduction to multiple-imputation analysis

[MI] **intro** — Introduction to mi

[MI] **Glossary**

Title

> **mi impute ologit** — Impute using ordered logistic regression

Syntax

> mi impute ologit *ivar* [*indepvars*] [*if*] [*weight*] [, *impute_options options*]

impute_options	description
Main	
* add(#)	specify number of imputations to add; required when no imputations exist
* replace	replace imputed values in existing imputations
rseed(#)	specify random-number seed
double	save imputed values in double precision; the default is to save them as float
Reporting	
dots	display dots as imputations are performed
noisily	display intermediate output
Advanced	
force	proceed with imputation even when missing imputed values are encountered
† noupdate	do not perform mi update; see [MI] **noupdate option**

* add(#) is required when no imputations exist; add(#) or replace is required if imputations exist.
† noupdate does not appear in the dialog box.

options	description
Main	
offset(*varname*)	include *varname* in model with coefficient constrained to 1
Maximization	
maximize_options	control the maximization process; seldom used

You must mi set your data before using mi impute ologit; see [MI] **mi set**.
fweights, iweights, and pweights are allowed; see [U] **11.1.6 weight**.

Menu

Statistics > Multiple imputation

Description

mi impute ologit fills in missing values of an ordinal variable using an ordered logistic regression imputation model. You can perform separate imputations on different subsets of the data by specifying an if condition. You can also account for frequency and sampling weights.

Options

    ◰ Main ◳

add(), replace, rseed(), double; see [MI] **mi impute**.

offset(*varname*); see [R] **estimation options**.

    ◰ Reporting ◳

dots, noisily; see [MI] **mi impute**. noisily specifies that the output from the ordered logistic regression fit to the observed data be displayed.

    ◰ Maximization ◳

maximize_options; see [R] **ologit**. These options are seldom used.

    ◰ Advanced ◳

force; see [MI] **mi impute**.

noupdate; see [MI] **noupdate option**.

Remarks

Remarks are presented under the following headings:

>Univariate imputation using ordered logistic regression
>Using mi impute ologit

See [MI] **mi impute** for a general description and details about options common to all imputation methods. Also see [MI] **workflow** for general advice on working with mi.

Univariate imputation using ordered logistic regression

The ordered logistic regression imputation method can be used to fill in missing values of an ordinal variable (e.g., Raghunathan et al. [2001] and van Buuren [2007]). It is a parametric method that assumes an underlying logistic model for the imputed variable. Similarly to the logistic imputation method, this method is based on the asymptotic approximation of the posterior predictive distribution of the missing data.

Using mi impute ologit

Following the example from [MI] **mi impute mlogit**, we consider the heart attack data (e.g., [MI] **intro substantive**, [MI] **mi impute**), where a logistic model of interest now includes information about alcohol consumption, variable alcohol—logit attack smokes age bmi female hsgrad i.alcohol.

```
. use http://www.stata-press.com/data/r11/mheart4
(Fictional heart attack data; alcohol missing)

. tabulate alcohol, missing
```

Alcohol consumption: none, <2 drinks/day, >=2 drinks/day	Freq.	Percent	Cum.
Do not drink	18	11.69	11.69
Less than 3 drinks/day	83	53.90	65.58
Three or more drinks/day	44	28.57	94.16
.	9	5.84	100.00
Total	154	100.00	

From the output, the alcohol variable has three unique ordered categories and 9 missing observations. We use the ordered logistic imputation method to impute missing values of alcohol. We create 10 imputations by specifying the add(10) option:

```
. mi set mlong

. mi register imputed alcohol
(9 m=0 obs. now marked as incomplete)

. mi impute ologit alcohol attack smokes age bmi female hsgrad, add(10)
Univariate imputation              Imputations =      10
Ordered logistic regression              added =      10
Imputed: m=1 through m=10              updated =       0
```

	Observations per m			
Variable	complete	incomplete	imputed	total
alcohol	145	9	9	154

(complete + incomplete = total; imputed is the minimum across m of the number of filled in observations.)

We can now analyze these multiply imputed data using logistic regression via **mi estimate**:

```
. mi estimate: logit attack smokes age bmi female hsgrad i.alcohol
  (output omitted)
```

(Continued on next page)

Saved results

`mi impute ologit` saves the following in `r()`:

Scalars
    `r(M)`   total number of imputations
    `r(M_add)`   number of added imputations
    `r(M_update)`   number of updated imputations
    `r(k_ivars)`   number of imputed variables (always one)

Macros
    `r(method)`   name of imputation method (ologit)
    `r(ivars)`   names of imputation variables
    `r(rseed)`   random-number seed

Matrices
    `r(N)`   number of observations in imputation sample
    `r(N_complete)`   number of complete observations in imputation sample
    `r(N_incomplete)`   number of incomplete observations in imputation sample
    `r(N_imputed)`   number of imputed observations in imputation sample

Methods and formulas

Consider a univariate variable $\mathbf{x} = (x_1, x_2, \ldots, x_n)'$ that contains K ordered categories and follows an ordered logistic model

$$\Pr(x_i = k | \mathbf{z}_i) = \Pr(\gamma_{k-1} < \mathbf{z}_i' \boldsymbol{\beta} + u \leq \gamma_k)$$
$$= \frac{1}{1 + \exp(-\gamma_k + \mathbf{z}_i'\boldsymbol{\beta})} - \frac{1}{1 + \exp(-\gamma_{k-1} + \mathbf{z}_i'\boldsymbol{\beta})} \qquad (1)$$

where $\mathbf{z}_i = (z_{i1}, z_{i2}, \ldots, z_{iq})'$ records values of predictors of $\mathbf{x}$ for observation i, $\boldsymbol{\beta}$ is the $q \times 1$ vector of unknown regression coefficients, and $\boldsymbol{\gamma} = (\gamma_1, \ldots, \gamma_{K-1})'$ are the unknown cutpoints with $\gamma_0 = -\infty$ and $\gamma_K = \infty$. (Note that there is no constant in this model because its effect is absorbed into the cutpoints; see [R] **ologit** for details.)

$\mathbf{x}$ contains missing values that are to be filled in. Consider the partition of $\mathbf{x} = (\mathbf{x}_o', \mathbf{x}_m')$ into $n_0 \times 1$ and $n_1 \times 1$ vectors containing the complete and the incomplete observations. Consider a similar partition of $\mathbf{Z} = (\mathbf{Z}_o, \mathbf{Z}_m)$ into $n_0 \times q$ and $n_1 \times q$ submatrices.

`mi impute ologit` follows the steps below to fill in $\mathbf{x}_m$:

1. Fit an ordered logistic model (1) to the observed data $(\mathbf{x}_o, \mathbf{Z}_o)$ to obtain the maximum likelihood estimates, $\widehat{\boldsymbol{\theta}} = (\widehat{\boldsymbol{\beta}}', \widehat{\boldsymbol{\gamma}}')'$, and their asymptotic sampling variance, $\widehat{\mathbf{U}}$.

2. Simulate new parameters, $\boldsymbol{\theta}_\star$, from the large-sample normal approximation, $N(\widehat{\boldsymbol{\theta}}, \widehat{\mathbf{U}})$, to its posterior distribution assuming the noninformative prior $\Pr(\boldsymbol{\theta}) \propto \text{const}$.

3. Obtain one set of imputed values, $\mathbf{x}_m^1$, by simulating from an ordered logistic distribution as defined by (1): one of K categories is randomly assigned to a missing category, i_m, using the cumulative probabilities computed from (1) with $\boldsymbol{\beta} = \boldsymbol{\beta}_\star$, $\boldsymbol{\gamma} = \boldsymbol{\gamma}_\star$, and $\mathbf{z}_i = \mathbf{z}_{i_m}$.

4. Repeat steps 2 and 3 to obtain M sets of imputed values, $\mathbf{x}_m^1, \mathbf{x}_m^2, \ldots, \mathbf{x}_m^M$.

Steps 2 and 3 above correspond to only approximate draws from the posterior predictive distribution of the missing data, $\Pr(\mathbf{x}_m | \mathbf{x}_o, \mathbf{Z}_o)$, because $\boldsymbol{\theta}_\star$ is drawn from the asymptotic approximation to its posterior distribution.

If weights are specified, a weighted ordered logistic regression model is fit to the observed data in step 1 (see [R] **ologit** for details).

References

Raghunathan, T. E., J. M. Lepkowski, J. Van Hoewyk, and P. Solenberger. 2001. A multivariate technique for multiply imputing missing values using a sequence of regression models. *Survey Methodology* 27: 85–95.

van Buuren, S. 2007. Multiple imputation of discrete and continuous data by fully conditional specification. *Statistical Methods in Medical Research* 16: 219–242.

Also see

[MI] **mi impute** — Impute missing values

[MI] **mi impute mlogit** — Impute using multinomial logistic regression

[MI] **mi estimate** — Estimation using multiple imputations

[MI] **intro** — Introduction to mi

[MI] **intro substantive** — Introduction to multiple-imputation analysis

Title

mi impute pmm — Impute using predictive mean matching

Syntax

mi impute pmm *ivar* [*indepvars*] [*if*] [*weight*] [, *impute_options options*]

impute_options	description
Main	
*add(#)	specify number of imputations to add; required when no imputations exist
*replace	replace imputed values in existing imputations
rseed(#)	specify random-number seed
double	save imputed values in double precision; the default is to save them as float
Reporting	
dots	display dots as imputations are performed
noisily	display intermediate output
Advanced	
force	proceed with imputation even when missing imputed values are encountered
†noupdate	do not perform mi update; see [MI] **noupdate option**

*add(#) is required when no imputations exist; add(#) or replace is required if imputations exist.

†noupdate does not appear in the dialog box.

options	description
Main	
noconstant	suppress constant term
knn(#)	specify # of closest observations (nearest neighbors) to draw from; default is knn(1)

You must mi set your data before using mi impute pmm; see [MI] **mi set**.

fweights, aweights, iweights, and pweights are allowed; see [U] **11.1.6 weight**.

Menu

Statistics > Multiple imputation

Description

mi impute pmm fills in missing values of a continuous variable using the predictive mean matching imputation method. You can perform separate imputations on different subsets of the data by specifying an if condition. You can also account for frequency, analytic, importance, and sampling weights.

Options

___Main___

noconstant; see [R] **estimation options**.

add(), replace, rseed(), double; see [MI] **mi impute**.

knn(#) specifies the number of closest observations (nearest neighbors) from which to draw imputed values. The default is to replace a missing value with the "closest" observation, knn(1). The closeness is determined based on the absolute difference between the linear prediction for the missing value and that for the complete values. The closest observation is the observation with the smallest difference. This option regulates the correlation among multiple imputations that affects the bias and the variability of the resulting multiple-imputation point estimates; see *Remarks* for details.

___Reporting___

dots, noisily; see [MI] **mi impute**. noisily specifies that the output from the linear regression fit to the observed data be displayed.

___Advanced___

force; see [MI] **mi impute**.

noupdate; see [MI] **noupdate option**.

Remarks

Remarks are presented under the following headings:

> *Univariate imputation using predictive mean matching*
> *Using mi impute pmm*

See [MI] **mi impute** for a general description and details about options common to all imputation methods. Also see [MI] **workflow** for general advice on working with mi.

Univariate imputation using predictive mean matching

Either predictive mean matching (pmm) or normal linear regression (regress) imputation methods can be used to fill in missing values of a continuous variable (Rubin 1987; Schenker and Taylor 1996). Predictive mean matching may be preferable to linear regression when the normality of the underlying model is suspect.

Predictive mean matching (PMM) is a partially parametric method that matches the missing value to the observed value with the closest predicted mean (or linear prediction). It was introduced by Little (1988) based on Rubin's (1986) ideas applied to statistical file matching. PMM combines the standard linear regression and the nearest-neighbor imputation approaches. It uses the normal linear regression to obtain linear predictions. It then uses the linear prediction, as a distance measure, to form the set of nearest neighbors (possible donors) consisting of the complete values. Finally, it randomly draws an imputed value from this set. By drawing from the observed data, PMM preserves the distribution of the observed values in the missing part of the data, which makes it more robust than the fully parametric linear regression approach.

With PMM, you need to decide how many nearest neighbors to include in the set of possible donors. `mi impute pmm` defaults to one nearest neighbor, `knn(1)`. You may need to include more depending on your data. The number of nearest neighbors affects the correlation among imputations—the smaller the number, the higher the correlation. High correlation in turn increases the variability of the MI point estimates. Including too many possible donors may result in increased bias of the MI point estimates. Thus the number of nearest neighbors regulates the tradeoff between the bias and the variance of the point estimators in repeated sampling. The literature does not provide a definitive recommendation on how to choose this number in practice; see Schenker and Taylor (1996) for some insight into this issue.

Using mi impute pmm

Recall the heart attack data from *Univariate imputation* in [MI] **mi impute**. We wish to fit a logistic regression of `attack` on some predictors, one of which, `bmi`, has missing values. To avoid losing information contained in complete observations of the other predictors, we impute `bmi`.

We showed one way of imputing `bmi` in [MI] **mi impute regress**. Suppose, however, that we want to restrict the imputed values of `bmi` to be within the range observed for `bmi`. We can use the PMM imputation method to achieve this. This method may also be preferable to the regression imputation of `bmi` because the distribution of `bmi` is slightly skewed.

```
. use http://www.stata-press.com/data/r11/mheart0
(Fictional heart attack data; bmi missing)
. mi set mlong
. mi register imputed bmi
(22 m=0 obs. now marked as incomplete)
. mi impute pmm bmi attack smokes age hsgrad female, add(20)
Univariate imputation               Imputations  =       20
Predictive mean matching                  added  =       20
Imputed: m=1 through m=20               updated  =        0
```

	Observations per m			
Variable	complete	incomplete	imputed	total
bmi	132	22	22	154

(complete + incomplete = total; imputed is the minimum across m of the number of filled in observations.)

By default, `mi impute pmm` uses one nearest neighbor to draw from. That is, it replaces missing values with an observed value whose linear prediction is the closest to that of the missing value. Using only one nearest neighbor may result in high variability of the MI estimates. You can increase the number of nearest neighbors from which the imputed value is drawn by specifying the `knn()` option. For example, we use 5 below:

```
. mi impute pmm bmi attack smokes age hsgrad female, replace knn(5)
Univariate imputation               Imputations  =       20
Predictive mean matching                  added  =        0
Imputed: m=1 through m=20               updated  =       20
```

	Observations per m			
Variable	complete	incomplete	imputed	total
bmi	132	22	22	154

(complete + incomplete = total; imputed is the minimum across m of the number of filled in observations.)

You can now refit the logistic model and examine the effect of using more neighbors:

```
. mi estimate: logit attack smokes age bmi hsgrad female
```
(output omitted)

See [MI] **mi impute**, [MI] **mi impute regress**, and [MI] **mi estimate** for more details.

Saved results

`mi impute pmm` saves the following in `r()`:

Scalars
- `r(M)`   total number of imputations
- `r(M_add)`   number of added imputations
- `r(M_update)`   number of updated imputations
- `r(knn)`   number of k nearest neighbors
- `r(k_ivars)`   number of imputed variables (always one)

Macros
- `r(method)`   name of imputation method (pmm)
- `r(ivars)`   names of imputation variables
- `r(rseed)`   random-number seed

Matrices
- `r(N)`   number of observations in imputation sample
- `r(N_complete)`   number of complete observations in imputation sample
- `r(N_incomplete)`   number of incomplete observations in imputation sample
- `r(N_imputed)`   number of imputed observations in imputation sample

Methods and formulas

`mi impute pmm` follows the steps as described in *Methods and formulas* of [MI] **mi impute regress** with the exception of step 3.

Consider a univariate variable $\mathbf{x} = (x_1, x_2, \ldots, x_n)'$ that follows a normal linear regression model

$$x_i | \mathbf{z}_i \sim N(\mathbf{z}_i' \boldsymbol{\beta}, \sigma^2) \qquad (1)$$

where $\mathbf{z}_i = (z_{i1}, z_{i2}, \ldots, z_{iq})'$ records values of predictors of $\mathbf{x}$ for observation i, $\boldsymbol{\beta}$ is the $q \times 1$ vector of unknown regression coefficients, and σ^2 is the unknown scalar variance. (Note that when a constant is included in the model—the default—$z_{i1} = 1$, $i = 1, \ldots, n$.)

$\mathbf{x}$ contains missing values that are to be filled in. Consider the partition of $\mathbf{x} = (\mathbf{x}_o', \mathbf{x}_m')'$ into $n_0 \times 1$ and $n_1 \times 1$ vectors containing the complete and the incomplete observations. Consider a similar partition of $\mathbf{Z} = (\mathbf{Z}_o, \mathbf{Z}_m)$ into $n_0 \times q$ and $n_1 \times q$ submatrices.

`mi impute pmm` follows the steps below to fill in $\mathbf{x}_m$ (for simplicity, we omit the conditioning on the observed data in what follows):

1. Fit a regression model (1) to the observed data $(\mathbf{x}_o, \mathbf{Z}_o)$ to obtain estimates $\widehat{\boldsymbol{\beta}}$ and $\widehat{\sigma}^2$ of the model parameters.

2. Simulate new parameters $\boldsymbol{\beta}_\star$ and $\sigma_\star^2$ from their joint posterior distribution under the conventional noninformative improper prior $\Pr(\boldsymbol{\beta}, \sigma^2) \propto 1/\sigma^2$. This is done in two steps:

$$\sigma_\star^2 \sim \widehat{\sigma}^2(n_0 - q)/\chi^2_{n_0-q}$$
$$\beta_\star|\sigma_\star^2 \sim N\left\{\widehat{\beta}, \sigma_\star^2(\mathbf{Z}_o'\mathbf{Z}_o)^{-1}\right\}$$

3. Generate the imputed values, $\mathbf{x}_m^1$, as follows. Let $\widehat{x}_i$ be the linear prediction of $\mathbf{x}$ based on predictors $\mathbf{Z}$ for observation i. Then for any missing observation i of $\mathbf{x}$, $x_i = x_{j_{\min}}$, where $j_{\min}$ is randomly drawn from the set of indices $\{i_1, i_2, \ldots, i_k\}$ corresponding to the first k minima determined based on the absolute differences between the linear prediction for incomplete observation i and linear predictions for all complete observations, $|\widehat{x}_i - \widehat{x}_j|$, $j \in obs$. For example, if $k = 1$ (the default), $j_{\min}$ is determined based on $|\widehat{x}_i - \widehat{x}_{j_{\min}}| = \min_{j \in \text{obs}} |\widehat{x}_i - \widehat{x}_j|$.

4. Repeat steps 2 and 3 to obtain M sets of imputed values, $\mathbf{x}_m^1, \mathbf{x}_m^2, \ldots, \mathbf{x}_m^M$.

If weights are specified, a weighted linear regression model is fit to the observed data in step 1 (see [R] **regress** for details).

References

Little, R. J. A. 1988. Missing-data adjustments in large surveys. *Journal of Business and Economic Statistics* 6: 287–296.

Rubin, D. B. 1986. Statistical matching using file concatenation with adjusted weights and multiple imputations. *Journal of Business and Economic Statistics* 4: 87–94.

———. 1987. *Multiple Imputation for Nonresponse in Surveys*. New York: Wiley.

Schenker, N., and J. M. G. Taylor. 1996. Partially parametric techniques for multiple imputation. *Computational Statistics & Data Analysis* 22: 425–446.

Also see

[MI] **mi impute** — Impute missing values

[MI] **mi impute regress** — Impute using linear regression

[MI] **mi estimate** — Estimation using multiple imputations

[MI] **intro** — Introduction to mi

[MI] **intro substantive** — Introduction to multiple-imputation analysis

Title

> **mi impute regress** — Impute using linear regression

Syntax

mi impute regress *ivar* [*indepvars*] [*if*] [*weight*] [, *impute_options options*]

impute_options	description
Main	
*add(#)	specify number of imputations to add; required when no imputations exist
*replace	replace imputed values in existing imputations
rseed(#)	specify random-number seed
double	save imputed values in double precision; the default is to save them as float
Reporting	
dots	display dots as imputations are performed
noisily	display intermediate output
Advanced	
force	proceed with imputation even when missing imputed values are encountered
†noupdate	do not perform mi update; see [MI] **noupdate option**

*add(#) is required when no imputations exist; add(#) or replace is required if imputations exist.

†noupdate does not appear in the dialog box.

options	description
Main	
noconstant	suppress constant term

You must mi set your data before using mi impute regress; see [MI] **mi set**.

fweights, aweights, iweights, and pweights are allowed; see [U] **11.1.6 weight**.

Menu

Statistics > Multiple imputation

Description

mi impute regress fills in missing values of a continuous variable using the Gaussian normal regression imputation method. You can perform separate imputations on different subsets of the data by specifying an if condition. You can also account for frequency, analytic, importance, and sampling weights.

181

Options

⌐ Main ⌐

noconstant; see [R] **estimation options**.

add(), replace, rseed(), double; see [MI] **mi impute**.

⌐ Reporting ⌐

dots, noisily; see [MI] **mi impute**. noisily specifies that the output from a linear regression fit to the observed data be displayed.

⌐ Advanced ⌐

force; see [MI] **mi impute**.

noupdate; see [MI] **noupdate option**.

Remarks

Remarks are presented under the following headings:

> *Univariate imputation using linear regression*
> *Using mi impute regress*

See [MI] **mi impute** for a general description and details about options common to all imputation methods. Also see [MI] **workflow** for general advice on working with mi.

Univariate imputation using linear regression

When a continuous variable contains missing values, a linear regression imputation method (or predictive mean matching; see [MI] **mi impute pmm**) can be used to fill in missing values (Rubin 1987; Schenker and Taylor 1996). The linear regression method is a fully parametric imputation method that relies on the normality of the model. Thus the imputation variable may need to be transformed from the original scale to meet the normality assumption prior to using mi impute regress.

The linear regression method is perhaps the most popular method for imputing quantitative variables. It is superior to other imputation methods when the underlying normal model holds. However, it can be more sensitive to violations of this assumption than other nonparametric and partially parametric imputation methods such as predictive mean matching. For example, Schenker and Taylor (1996) studied the sensitivity of the regression method to the misspecification of the regression function and error distribution. They found that this method still performs well in the presence of heteroskedasticity and when the error distribution is heavier-tailed than the normal. However, it resulted in increased bias and variances under a misspecified regression function.

Using mi impute regress

Recall the heart attack data from *Univariate imputation* of [MI] **mi impute**. We wish to fit a logistic regression of attack on some predictors, one of which, bmi, has missing values. To avoid losing information contained in complete observations of the other predictors, we impute bmi.

The distribution of BMI is slightly skewed to the right and so we choose to fill in missing values of BMI on a log-transformed scale here. To do that, we need to create a new variable, lnbmi, containing the log of bmi and impute it:

```
. use http://www.stata-press.com/data/r11/mheart0
(Fictional heart attack data; bmi missing)

. generate lnbmi = ln(bmi)
(22 missing values generated)

. mi set mlong

. mi register imputed lnbmi
(22 m=0 obs. now marked as incomplete)
```

Following the steps in *Imputing transformations of imputed variables* of [MI] **mi impute**, we create imputed variable lnbmi containing the log of bmi and register it as imputed. We omitted the step of eliminating possible ineligible missing values in lnbmi because bmi ranges from 17 to 38 and we do not anticipate any extra (algebraic) missing from the operation ln(bmi).

We now use mi impute to impute missing values of lnbmi. We create 20 imputations and specify a random-number seed for reproducibility:

```
. mi impute regress lnbmi age attack smokes age hsgrad female, add(20) rseed(2232)
Univariate imputation              Imputations =      20
Linear regression                        added =      20
Imputed: m=1 through m=20              updated =       0
```

	Observations per m			
Variable	complete	incomplete	imputed	total
lnbmi	132	22	22	154

(complete + incomplete = total; imputed is the minimum across m of the number of filled in observations.)

From the output, all 22 incomplete values of lnbmi are imputed.

We want to use BMI in its original scale in the analysis. To do that we need to replace bmi with exponentiated lnbmi. Because bmi now is a function of the imputed variable, it becomes a passive variable:

```
. mi register passive bmi
. quietly mi passive: replace bmi = exp(lnbmi)
```

(*Continued on next page*)

Finally, we fit the logistic regression:

```
. mi estimate, dots: logit attack smokes age bmi hsgrad female
Imputations (20):
.........10.........20 done

Multiple-imputation estimates            Imputations     =         20
Logistic regression                      Number of obs   =        154
                                         Average RVI     =     0.0252
DF adjustment:     Large sample          DF:      min    =    1752.87
                                                  avg    =  238565.50
                                                  max    =  605106.21
Model F test:      Equal FMI             F(   5,111161.3)=       3.69
Within VCE type:        OIM              Prob > F        =     0.0024
```

attack	Coef.	Std. Err.	t	P>|t|	[95% Conf. Interval]	
smokes	1.211756	.3593305	3.37	0.001	.5074783	1.916035
age	.0361334	.0154753	2.33	0.020	.0058023	.0664645
bmi	.1097576	.0470854	2.33	0.020	.017408	.2021071
hsgrad	.1649715	.4055386	0.41	0.684	-.6298738	.9598168
female	-.1030983	.4174652	-0.25	0.805	-.921318	.7151214
_cons	-5.642962	1.674039	-3.37	0.001	-8.924846	-2.361077

We obtain results comparable with those from [MI] **intro substantive**.

Saved results

mi impute regress saves the following in r():

Scalars
    r(M)   total number of imputations
    r(M_add)   number of added imputations
    r(M_update)   number of updated imputations
    r(k_ivars)   number of imputed variables (always one)

Macros
    r(method)   name of imputation method (regress)
    r(ivars)   names of imputation variables
    r(rseed)   random-number seed

Matrices
    r(N)   number of observations in imputation sample
    r(N_complete)   number of complete observations in imputation sample
    r(N_incomplete)   number of incomplete observations in imputation sample
    r(N_imputed)   number of imputed observations in imputation sample

Methods and formulas

Consider a univariate variable $\mathbf{x} = (x_1, x_2, \ldots, x_n)'$ that follows a normal linear regression model

$$x_i | \mathbf{z}_i \sim N(\mathbf{z}_i' \boldsymbol{\beta}, \sigma^2) \tag{1}$$

where $\mathbf{z}_i = (z_{i1}, z_{i2}, \ldots, z_{iq})'$ records values of predictors of $\mathbf{x}$ for observation i, $\boldsymbol{\beta}$ is the $q \times 1$ vector of unknown regression coefficients, and σ^2 is the unknown scalar variance. (Note that when a constant is included in the model—the default—$z_{i1} = 1$, $i = 1, \ldots, n$.)

$\mathbf{x}$ contains missing values that are to be filled in. Consider the partition of $\mathbf{x} = (\mathbf{x}'_o, \mathbf{x}'_m)$ into $n_0 \times 1$ and $n_1 \times 1$ vectors containing the complete and the incomplete observations. Consider a similar partition of $\mathbf{Z} = (\mathbf{Z}_o, \mathbf{Z}_m)$ into $n_0 \times q$ and $n_1 \times q$ submatrices.

mi impute regress follows the steps below to fill in $\mathbf{x}_m$ (for simplicity, we omit the conditioning on the observed data in what follows):

1. Fit a regression model (1) to the observed data $(\mathbf{x}_o, \mathbf{Z}_o)$ to obtain estimates $\widehat{\boldsymbol{\beta}}$ and $\widehat{\sigma}^2$ of the model parameters.

2. Simulate new parameters $\boldsymbol{\beta}_\star$ and $\sigma^2_\star$ from their joint posterior distribution under the conventional noninformative improper prior $\Pr(\boldsymbol{\beta}, \sigma^2) \propto 1/\sigma^2$. This is done in two steps:

$$\sigma^2_\star \sim \widehat{\sigma}^2(n_0 - q)/\chi^2_{n_0 - q}$$
$$\boldsymbol{\beta}_\star | \sigma^2_\star \sim N\left\{\widehat{\boldsymbol{\beta}}, \sigma^2_\star(\mathbf{Z}'_o \mathbf{Z}_o)^{-1}\right\}$$

3. Obtain one set of imputed values, $\mathbf{x}^1_m$, by simulating from $N(\mathbf{Z}_m \boldsymbol{\beta}_\star, \sigma^2_\star I_{n_1 \times n_1})$.

4. Repeat steps 2 and 3 to obtain M sets of imputed values, $\mathbf{x}^1_m, \mathbf{x}^2_m, \ldots, \mathbf{x}^M_m$.

Steps 2 and 3 above correspond to simulating from the posterior predictive distribution of the missing data $\Pr(\mathbf{x}_m | \mathbf{x}_o, \mathbf{Z}_o)$ (for example, see Gelman et al. [2004, 355–358]).

If weights are specified, a weighted linear regression model is fit to the observed data in step 1 (see [R] **regress** for details). Also, in the case of aweights, $\sigma^2_\star I_{n_1 \times n_1}$ is replaced with $\sigma^2_\star \mathbf{W}^{-1}_{n_1 \times n_1}$ in step 3, where $\mathbf{W} = \text{diag}\{w_i\}$ and w_i is the analytic weight for observation i.

References

Gelman, A., J. B. Carlin, H. S. Stern, and D. B. Rubin. 2004. *Bayesian Data Analysis*. 2nd ed. London: Chapman & Hall/CRC.

Rubin, D. B. 1987. *Multiple Imputation for Nonresponse in Surveys*. New York: Wiley.

Schenker, N., and J. M. G. Taylor. 1996. Partially parametric techniques for multiple imputation. *Computational Statistics & Data Analysis* 22: 425–446.

Also see

[MI] **mi impute** — Impute missing values

[MI] **mi impute pmm** — Impute using predictive mean matching

[MI] **mi estimate** — Estimation using multiple imputations

[MI] **intro** — Introduction to mi

[MI] **intro substantive** — Introduction to multiple-imputation analysis

Title

mi merge — Merge mi data

Syntax

    mi merge 1:1 *varlist* using *filename* [, *options*]

    mi merge m:1 *varlist* using *filename* [, *options*]

    mi merge 1:m *varlist* using *filename* [, *options*]

    mi merge m:m *varlist* using *filename* [, *options*]

options	description
Main	
<u>g</u>enerate(*newvar*)	create *newvar* recording how observations matched
<u>nol</u>abel	do not copy value-label definitions from using
<u>non</u>otes	do not copy notes from using
force	allow string/numeric variable type mismatch without error
Results	
assert(*results*)	require observations to match as specified
keep(*results*)	results to keep
<u>nore</u>port	do not display result summary table
<u>noup</u>date	see [MI] **noupdate option**

Notes:

1. Jargon:
   match variables = *varlist*, variables on which match performed
         master = data in memory
         using = data on disk (*filename*)

2. Master must be mi set; using may be mi set.

3. mi merge is syntactically and logically equivalent to merge (see [D] **merge**).

4. mi merge syntactically differs from merge in that the nogenerate, sorted, keepusing(), update, and replace options are not allowed. Also, no _merge variable is created unless the generate() option is specified.

5. *filename* must be enclosed in double quotes if *filename* contains blanks or other special characters.

Menu

Statistics > Multiple imputation

Description

`mi merge` is `merge` for `mi` data; see [D] **merge** for a description of merging datasets.

It is recommended that the match variables (*varlist* in the syntax diagram) not include imputed or passive variables, or any varying or super-varying variables. If they do, the values of the match variables in $m=0$ will be used to control the merge even in $m=1$, $m=2$, ..., $m=M$. Thus $m=0, m=1, \ldots, m=M$ will all be merged identically, and there will continue to be a one-to-one correspondence between the observations in $m=0$ with the observations in each of $m>0$.

Options

Main

generate(*newvar*) creates new variable *newvar* containing the match status of each observation in the resulting data. The codes are 1, 2, and 3 from the table above.

nolabel prevents copying the value-label definitions from the using data to the master. Even if you do not specify this option, label definitions from the using never replace those of the master.

nonotes prevents any notes in the using from being incorporated into the master; see [D] **notes**.

force allows string/numeric variable type mismatches, resulting in missing values from the using dataset. If omitted, `mi merge` issues an error; if specified, `mi merge` issues a warning.

Results

assert(*results*) specifies how observations should match. If results are not as you expect, an error message will be issued and the master data left unchanged.

code	word	description
1	master	observation appeared in master only
2	using	observation appeared in using only
3	match	observation appeared in both

(Numeric codes and words are equivalent; you may use either.)

assert(match) specifies that all observations in both the master and the using are expected to match, and if that is not so, an error message is to be issued. assert(match master) means that all observations match or originally appeared only in the master. See [D] **merge** for more information.

keep(*results*) specifies which observations are to be kept from the merged dataset. keep(match) would specify that only matches are to be kept.

noreport suppresses the report that `mi merge` ordinarily presents.

noupdate in some cases suppresses the automatic `mi update` this command might perform; see [MI] **noupdate option**.

Remarks

Use `mi merge` when you would use `merge` if the data were not `mi`.

Remarks are presented under the following headings:

Merging with non-mi data
Merging with mi data
Merging with mi data containing overlapping variables

Merging with non-mi data

Assume that file `ipats.dta` contains data on the patients in the ICU of a local hospital. The data are `mi set`, $M = 5$, and missing values have been imputed. File `nurses.dta` contains information on nurses and is not `mi` data. You wish to add the relevant nurse information to each patient. Type

```
. use ipats, clear
. mi merge m:1 nurseid using nurses, keep(master)
```

The resulting data are still `mi set` with $M = 5$. The new variables are unregistered.

Merging with mi data

Now assume the same situation as above, except this time `nurses.dta` is `mi` data. Some of the nurse variables have missing values, and those values have been imputed. M is 6. To combine the datasets, you type the same as you would have typed before:

```
. use ipats, clear
. mi merge m:1 nurseid using nurses, keep(master)
```

Remember, $M = 5$ in `ipats.dta` and $M = 6$ in `nurses.dta`. The resulting data have $M = 6$, the larger value. There are missing values in the patient variables in $m = 6$, so we need to either impute them or drop the extra imputation by typing `mi set M = 5`.

Merging with mi data containing overlapping variables

Now assume the situation as directly above, but this time `nurses.dta` contains variables other than `nurseid` that also appear in `ipats.dta`. Such variables—variables in common that are not used as matching variables—are called overlapping variables. Assume `seniornurse` is such a variable. Let's imagine that `seniornurse` has no missing values and is unregistered in `ipats.dta`, but does have missing values and is registered as imputed in `nurses.dta`.

You will want `seniornurse` registered as imputed if merging `nurses.dta` adds new observations that have `seniornurse` equal to missing. On the other hand, if none of the added observations has `seniornurse` equal to missing, then you will want the variable left unregistered. And that is exactly what `mi merge` does. That is,

- Variables unique to the master will be registered according to how they were registered in the master.

- Variables unique to the using will be registered according to how they were registered in the using.

- Variables that overlap will be registered according to how they were in the master if there are no unmatched using observations in the final result.

- If there are such unmatched using observations in the final result, then the unique variables that do not contain missing in the unmatched-and-kept observations will be registered according to how they were registered in the master. So will all variables registered as imputed in the master.

- Variables that do contain missing in the unmatched-and-kept observations will be registered as imputed if they were registered as imputed in the using data, or as passive if they were registered as passive in the using data.

Thus variables might be registered differently if we typed

    . mi merge m:1 nurseid using nurses, keep(master)

rather than

    . mi merge m:1 nurseid using nurses, gen(howmatch)
    . keep if howmatch==3

If you want to keep the matched observations, it is better to specify merge's keep() option.

Saved results

mi merge saves the following in r():

Scalars
    r(N_master) number of observations in $m=0$ in master
    r(N_using)   number of observations in $m=0$ in using
    r(N_result) number of observations in $m=0$ in result

    r(M_master) number of imputations (M) in master
    r(M_using)   number of imputations (M) in using
    r(M_result) number of imputations (M) in result

Macros
    r(newvars)   new variables added

Thus values in the resulting data are

N = # of observations in $m = 0$
   = r(N_result)

k = # of variables
  = k_master + ':word count 'r(newvars)''

M = # of imputations
    = max(r(M_master), r(M_using))
    = r(M_result)

Also see

[MI] **intro** — Introduction to mi

[D] **merge** — Merge datasets

[MI] **mi append** — Append mi data

Title

> **mi misstable** — Tabulate pattern of missing values

Syntax

    mi <u>misstab</u>le <u>sum</u>marize [*varlist*] [*if*] [, *options*]

    mi <u>misstab</u>le <u>pat</u>terns [*varlist*] [*if*] [, *options*]

    mi <u>misstab</u>le tree [*varlist*] [*if*] [, *options*]

    mi <u>misstab</u>le <u>nest</u>ed [*varlist*] [*if*] [, *options*]

options	description
Main	
exmiss	treat .a, .b, ..., .z as missing
m(#)	run misstable on $m = \#$; default is $m = 0$
other_options	see [R] **misstable** (exok is not allowed)
nopreserve	programmer's option; see [P] **nopreserve option**

Menu

Statistics > Multiple imputation

Description

    mi misstable runs misstable on $m = 0$, or on $m = \#$ if the m(#) option is specified. misstable makes tables to help in understanding the pattern of missing values in your data; see [R] **misstable**.

Options

       Main

  exmiss specifies that the extended missing values, .a, .b, ..., .z, are to be treated as missing. misstable treats them as missing by default and has the exok option to treat them as nonmissing. mi misstable turns that around and has the exmiss option.

  In the mi system, extended missing values that are recorded in imputed variables indicate values not to be imputed and thus are, in a sense, not missing, or more accurately, missing for a good and valid reason.

  The exmiss option is intended for use with the patterns, tree, and nested subcommands. You may specify exmiss with the summarize subcommand, but the option is ignored because summarize reports both extended and system missing in separate columns.

m(#) specifies the imputation dataset on which misstable is to be run. The default is $m = 0$, the original data.

other_options are allowed; see [R] **misstable**.

Remarks

See [R] **misstable**.

Saved results

See [R] **misstable**.

Also see

[MI] **intro** — Introduction to mi

[R] **misstable** — Tabulate missing values

[MI] **mi varying** — Identify variables that vary across imputations

Title

> **mi passive** — Generate/replace and register passive variables

Syntax

    mi passive: { generate | egen | replace } ...

    mi passive: by *varlist*: { generate | egen | replace } ...

The full syntax is

    mi passive [, *options*] : [by *varlist* [(*varlist*)] :] { generate | egen | replace } ...

options	description
noupdate	see [MI] **noupdate option**
nopreserve	do not first preserve

Also see [D] **generate** and [D] **egen**.

Menu

Statistics > Multiple imputation

Description

    mi passive creates and registers passive variables or replaces the contents of existing passive variables.

    More precisely, mi passive executes the specified generate, egen, or replace command on each of $m = 0$, $m = 1$, ..., $m = M$. If the command is generate or egen, then mi passive registers the new variable as passive. If the command is replace, then mi passive verifies that the variable is already registered as passive.

Options

    noupdate in some cases suppresses the automatic mi update this command might perform; see [MI] **noupdate option**.

    nopreserve is a programmer's option. It specifies that mi passive is not to preserve the data if it ordinarily would. This is used by programmers who have already preserved the data before calling mi passive.

Remarks

Remarks are presented under the following headings:

    mi passive basics
    mi passive works with the by prefix
    mi passive works fastest with the wide style
    Renaming passive variables
    Dropping passive variables
    Update passive variables when imputed values change
    Alternatives to mi passive

mi passive basics

A passive variable is a variable that is a function of imputed variables or of other passive variables. For instance, if variable age were imputed and you created lnage from it, the lnage variable would be passive. The right way to create lnage is to type

    . mi passive: generate lnage = ln(age)

Simply typing

    . generate lnage = ln(age)

is not sufficient because that would create lnage in the $m = 0$ data, and age, being imputed, varies across m. There are situations where omitting the mi passive prefix would be almost sufficient, namely, when the data are mlong or flong style, but even then you would need to follow up by typing mi register passive lnage.

To create passive variables or to change the values of existing passive variables, use mi passive.

mi passive works with the by prefix

You can use mi passive with the by prefix. For instance, you can type

    . mi passive: by person: generate totaltodate = sum(amount)

or

    . mi passive: by sex: egen avg = mean(income)

You do not need to sort the data before issuing either of these commands, nor are you required to specify by's sort option. mi passive handles sorting issues for you.

Use by's parenthetical syntax to specify the order within by, if that is necessary. For instance,

    . mi passive: by person (time): generate lastamount = amount[_n-1]

Do not omit the parenthetical time and instead attempt to sort the data yourself:

    . sort person time
    . mi passive: by person: generate lastamount = amount[_n-1]

Sorting the data yourself will work if your data happen to be wide style; it will not work in general.

mi passive works fastest with the wide style

mi passive works with any style, but it works fastest when the data are wide style. If you are going to issue multiple mi passive commands, you can usually speed execution by first converting your data to the wide style; see [MI] **mi convert**.

Renaming passive variables

Use mi rename (see [MI] **mi rename**) to rename all variables, not just passive variables:

. mi rename *oldname newname*

rename (see [D] **rename**) is insufficient for renaming passive variables regardless of the style of your data.

Dropping passive variables

Use drop (see [D] **drop**) to drop variables (or observations), but run mi update (see [MI] **mi update**) afterward.

. drop *var_or_vars*

. mi update

This advice applies for all variables, not just passive ones.

Update passive variables when imputed values change

Passive variables are not automatically updated when the values of the underlying imputed variables change.

If imputed values change, or if you add more imputations, you must update or re-create the passive variables. If you have several passive variables, we suggest you make a do-file to create them. You can run the do-file again whenever necessary. A do-file to create lnage and totaltodate might read

────────────────────────────────── begin cr_passive.do ──────────
```
use mydata, clear
capture drop lnage
capture drop totaltodate
mi update
mi passive: generate lnage = ln(age)
mi passive: by person (time): generate totaltodate = sum(amount)
```
────────────────────────────────── end cr_passive.do ──────────

Alternatives to mi passive

mi passive can run any generate, replace, or egen command. If that is not sufficient to create the variable you need, you will have to create the variable for yourself. Here is how you do that:

1. If your data are wide or mlong, use mi convert (see [MI] **mi convert**) to convert them to one of the fully long styles, flong or flongsep, and then continue with the appropriate step below.

2. If your data are flong, mi system variable _mi_m records m. Create your new variable by using standard Stata commands, but do that by _mi_m. After creating the variable, mi register it as passive; see [MI] **mi set**.

3. If your data are flongsep, create the new variable in each of the $m = 0$, $m = 1$, ..., $m = M$ datasets, and then register the result. Start by working with a copy of your data:

    . mi copy *newname*

    The data in memory at this point correspond to $m = 0$. Create the new variable and then save the data:

    . *(create new_variable)*

    . save *newname*, replace

    Now use the $m = 1$ data and repeat the process:

    . use _1_*newname*

    . *(create new_variable)*

    . save _1_*newname*, replace

    Repeat for $m = 2$, $m = 3$, ..., $m = M$.

    At this point, the new variable is created but not yet registered. Reload the original $m = 0$ data, register the new variable as passive, and run mi update (see [MI] **mi update**):

    . use *newname*

    . register passive *new_variable*

    . mi update

    Finally, copy the result back to your original flongsep data,

    . mi copy *name*, replace

    or if you started with mlong, flong, or wide data, then convert the data back to your preferred style:

    . mi convert *original_style*

    Either way, erase the *newname* flongsep dataset collection:

    . mi erase *newname*

The third procedure can be tedious and error-prone if M is large. We suggest that you make a do-file to create the variable and then run it on each of the $m = 0$, $m = 1$, ..., $m = M$ datasets:

```
. mi copy newname
. do mydofile
. save newname, replace
. forvalues m=1(1)20 {                  // we assume M=20
>     use _`m'_newname
>     do mydofile
>     save _`m'_newname, replace
> }
. use newname
. register passive new_variable
. mi update
```

Also see

[MI] **intro** — Introduction to mi

[MI] **mi reset** — Reset imputed or passive variables

[MI] **mi xeq** — Execute command(s) on individual imputations

Title

mi ptrace — Load parameter-trace file into Stata

Syntax

    mi ptrace describe [using] *filename*

    mi ptrace use *filename* [, *use_options*]

use_options	description
clear	okay to replace existing data in memory
double	load variables as doubles (default is floats)
select(*selections*)	what to load (default is all)

where *selections* is a space-separated list of individual selections. Individual selections are of the form

    b[*yname*, *xname*]
    v[*yname*, *yname*]

where *yname*s and *xname*s are displayed by `mi ptrace describe`. You can also specify

    b[#_y, #_x]
    v[#_y, #_y]

where #_y and #_x are the variable numbers associated with *yname* and *xname*, and those too are shown by `mi ptrace describe`.

For b, you can also specify * to mean all possible index elements. For instance,

b[*,*]	all elements of b
b[*yname*,*]	row corresponding to *yname*
b[*,*xname*]	column corresponding to *xname*

Similarly, b[#_y,*] and b[*,#_x] are allowed. The same is allowed for v, and also, the second element can be specified as <, <=, =, >=, or >. For instance,

v[*yname*,=]	variance of *yname*
v[*,=]	all variances (diagonal elements)
v[*,<]	lower triangle
v[*,<=]	lower triangle and diagonal
v[*,>=]	upper triangle and diagonal
v[*,>]	upper triangle

In `mi ptrace describe` and in `mi ptrace use`, *filename* must be specified in quotes if it contains special characters or blanks. *filename* is assumed to be *filename*.stptrace if the suffix is not specified.

Description

Parameter-trace files, files with suffix .stptrace, are created by the saveptrace() option of mi impute mvn; see [MI] **mi impute mvn**. These are not Stata datasets, but they can be loaded as if they were by using mi ptrace use. Their contents can be described without loading them by using mi ptrace describe.

Options

clear specifies that it is okay to clear the dataset in memory, even if it has not been saved to disk since it was last changed.

double specifies that elements of b and v are to be loaded as doubles; they are loaded as floats by default.

select(*selections*) allows you to load subsets of b and v. If the option is not specified, all of b and v are loaded. That result is equivalent to specifying select(b[*,*] v[*,<=]). The <= specifies that just the diagonal and lower triangle of symmetric matrix v be loaded.

Specifying select(b[*,*]) would load just b.

Specifying select(v[*,<=]) would load just v.

Specifying select(b[*,*] v[*,=]) would load b and the diagonal elements of v.

Remarks

Say that we impute the values of y_1 and y_2 assuming that they are multivariate normal distributed, with their means determined by a linear combination of x_1, x_2, and x_3, and their variance constant. Writing this more concisely, $\mathbf{y} = (y_1, y_2)'$ is distributed $\mathrm{MVN}(\mathbf{XB}, \mathbf{V})$, where $\mathbf{B}$: 2×3 and $\mathbf{V}$: 2×2. If we use MCMC or EM procedures to produce values of $\mathbf{B}$ and $\mathbf{V}$ to be used to generate values for $\mathbf{y}$, we must ensure that we use sufficient iterations so that the iterative procedure stabilizes. mi impute mvn provides the worst linear combination (WLC) of the elements of $\mathbf{B}$ and $\mathbf{V}$. If we want to perform other checks, we can specify mi impute mvn's saveptrace(*filename*) option. mi impute then produces a file containing m (imputation number), iter (overall iteration number), and the corresponding B and V. The last iter for each m is the B and V that mi impute mvn used to impute the missing values.

When we used mi impute mvn, we specified burn-in and burn-between numbers, say, burnin(300) and burnbetween(100). If we also specified saveptrace(), the file produced is organized as follows:

```
    record #  |    m      iter       B         V
           1  |    1      -299       ...       ...
           2  |    1      -298       ...       ...
           .  |    .       .          .         .
           .  |    .       .          .         .
         299  |    1       -1        ...       ...
         300  |    1        0        ...       ...   <- used to impute m=1
         301  |    2        1         .         .
         302  |    2        2         .         .
           .  |    .       .          .         .
           .  |    .       .          .         .
        399.  |    1       99        ...       ...
        400.  |    1      100        ...       ...   <- used to impute m=2
        401.  |    2      101        ...       ...
           .  |    .       .          .         .
           .  |    .       .          .         .
```

The file is not a Stata dataset, but `mi ptrace use` can load the file and convert it into Stata format, and then it will look just like the above, except for the following:

- The record number will become the Stata observation number.
- B will become variables b_y1x1, b_y1x2, and b_y1x3; and b_y2x1, b_y2x2, and b_y2x3. (Remember, we had 2 y variables and 3 x variables.)
- V will become variables v_y1y1, v_y2y1, and v_y2y2. (This is the diagonal and lower triangle of **V**; variable v_y1y2 is not created because it would be equal to v_y2y1.)
- Variable labels will be filled in with the underlying names of the variables. For instance, the variable label for b_y1x1 might be "experience, age", and that would remind us that b_y1x1 contains the coefficient on age used to predict experience. v_y2y1 might be "education, experience", and that would remind us that v_y2y1 contains the covariance between education and experience.

Saved results

`mi ptrace describe` saves the following in r():

Scalars
    r(tc)   %tc date-and-time file created
    r(nx)   number of x variables (columns of B)
    r(nx)   number of y variables (rows of B)

Macros
    r(x)   space-separated [op.]varname of x
    r(y)   space-separated [op.]varname of y
    r(id)   name of file creator

Also see

[MI] **intro** — Introduction to mi

[MI] **mi impute mvn** — Impute using multivariate normal regression

Title

mi rename — Rename variable

Syntax

mi <u>ren</u>ame *oldname newname* [, <u>noup</u>date]

Menu

Statistics > Multiple imputation

Description

mi rename is rename (see [D] **rename**) for mi data; it renames variables.

Option

noupdate in some cases suppresses the automatic mi update this command might perform; see [MI] **noupdate option**.

Remarks

Remarks are presented under the following headings:

> *Specifying the noupdate option*
> *What to do if you accidentally use rename*
> *What to do if you accidentally use rename on wide data*
> *What to do if you accidentally use rename on mlong data*
> *What to do if you accidentally use rename on flong data*
> *What to do if you accidentally use rename on flongsep data*

Specifying the noupdate option

If you are renaming more than one variable, you can speed execution with no loss of safety by specifying the noupdate option after the first mi rename:

```
. mi rename ageyears age
. mi rename timeinstudy studytime, noupdate
. mi rename personid id, noupdate
```

The above is generally good advice. When giving one mi command after another, you may specify noupdate after the first command to speed execution.

What to do if you accidentally use rename

Assume that you just typed

    . rename ageyears age

rather than typing

    . mi rename ageyears age

as you should have. No damage has been done yet, but if you give another mi command and it runs mi update, real damage will be done. We will discuss that and what to do about it in the sections that follow, but first, if you have given no additional mi commands, use rename (not mi rename) to rename the variable back to how it was:

    . rename age ageyears

Then use mi rename as you should have in the first place:

    . mi rename ageyears age

The sections below handle the case where mi update has run. You will know that mi update has run because since the rename, you gave some mi command—perhaps even mi update itself—and you saw a message like one of these:

> (variable ageyears dropped in $m > 0$)
>
> (imputed variable ageyears unregistered because not in $m = 0$)
>
> (passive variable ageyears unregistered because not in $m = 0$)
>
> (regular variable ageyears unregistered because not in $m = 0$)

What to do if you accidentally use rename on wide data

If ageyears was unregistered, no damage was done, and no additional action needs to be taken.

If ageyears was registered as regular, no damage was done. However, your renamed variable is no longer registered. Reregister the variable under its new name by typing mi register regular age.

If ageyears was registered as imputed or passive, you just lost all values for $m > 0$. Passive variables are usually not too difficult to re-create; see [MI] **mi passive**. If the variable was imputed, well, hope that you will have saved your data recently when you make this error and, before that, learn good computing habits.

What to do if you accidentally use rename on mlong data

If ageyears was unregistered, no damage was done, and no additional action needs to be taken.

If ageyears was registered as regular, no damage was done. However, your renamed variable is no longer registered. Reregister the variable under its new name by typing mi register regular age.

If ageyears was registered as imputed or passive, you just lost all values for $m > 0$. We offer the same advice as we offered when the data were wide: Passive variables are usually not too difficult to re-create—see [MI] **mi passive**—and otherwise hope that you will have saved your data recently when you make this error. It is always a good idea to save your data periodically.

What to do if you accidentally use rename on flong data

The news is better in this case; no matter how your variables were registered, you have not lost data.

If ageyears was unregistered, no further action is required.

If ageyears was registered as regular, you need to reregister the variable under its new name by typing mi register regular age.

If ageyears was registered as passive or imputed, you need to reregister the variable under its new name by typing mi register passive age or mi register imputed age.

What to do if you accidentally use rename on flongsep data

The news is not as good in this case.

If ageyears was unregistered, no damage was done. When mi update ran, it noticed that old variable ageyears no longer appeared in $m > 0$ and that new variable age now appeared in $m = 0$, so mi update dropped the first and added the second to $m > 0$, thus undoing any damage. There is nothing more that needs to be done.

If ageyears was registered as regular, no damage was done, but you need to reregister the variable by typing mi register regular age.

If ageyears was registered as passive or imputed, you have lost the values in $m > 0$. Now would probably be a good time for us to mention how you should work with a copy of your flongsep data; see [MI] **mi copy**.

Also see

[MI] **intro** — Introduction to mi

Title

mi replace0 — Replace original data

Syntax

mi replace0 using *filename*, id(*varlist*)

Typical use is

. mi extract 0
. (*perform data-management commands*)
. mi replace0 using *origfile*, id(*idvar*)

Menu

Statistics > Multiple imputation

Description

mi replace0 replaces $m = 0$ of an mi dataset with the non-mi data of another and then carries out whatever changes are necessary in $m > 0$ of the former to make the resulting mi data consistent. mi replace0 starts with one of the datasets in memory and the other on disk (it does not matter which is which) and leaves in memory the mi data with $m = 0$ replaced.

Option

id(*varlist*) is required; it specifies the variable or variables that can be used to match the observations in $m = 0$ of the mi data to the observations of the non-mi dataset. The ID variables must uniquely identify the observations in each dataset, and equal values across datasets must indicate corresponding observations, but one, the other, or both datasets can have observations found (or not found) in the other.

Remarks

It is often easier to perform data management on $m = 0$ and then let mi replace0 duplicate the results for $m = 1$, $m = 2$, ..., $m = M$ rather than perform the data management on all m's simultaneously. It is easier because $m = 0$ by itself is a non-mi dataset, so you can use any of the general Stata commands (that is, non-mi commands) with it.

You use mi extract to extract $m = 0$; see [MI] **mi extract**. The extracted dataset is just a regular Stata dataset; it is not mi set, nor does it have any secret complexities.

You use mi replace0 to recombine the datasets after you have modified the $m = 0$ data. mi replace0 can deal with the following changes to $m = 0$:

- changes to the values of existing variables,
- removal of variables,

- addition of new variables,
- dropped observations, and
- added observations.

For instance, you could use `mi extract` and `mi replace0` to do the following:

```
. use my_midata, clear
. mi extract 0
. replace age = 26 if age==6
. replace age = 32 if pid==2088
. merge 1:1 pid using newvars, keep(match) nogen
. by location: egen avgrate = mean(rate)
. drop proxyrate
. mi replace0 using my_midata, id(pid)
```

In the above,

1. we extract $m = 0$;
2. we update existing variable `age` (we fix a typo and, having finally obtained the age of `pid` 2088, add it);
3. we merge $m = 0$ with `newvars.dta` to obtain some new variables and, in the process, keep only the observations that were found in both $m = 0$ and `newvars.dta`;
4. we create new variable `avgrate` equal to the mean rate by location; and
5. we drop previously existing variable `proxyrate`.

We then take that result and use it to replace $m = 0$ in our original `mi` dataset. We leave it to `mi replace0` to carry out the changes to $m = 1$, $m = 2$, ..., $m = M$ to account for what we did to $m = 0$.

By the way, it turns out that `age` in `my_midata.dta` is registered as imputed. Note that we changed one nonmissing value to another nonmissing value and changed another missing value to a nonmissing value. `mi replace0` will deal with the implications of that. It would even deal with us having changed a nonmissing value to a missing value.

There is no easier way to do data management than by using `mi extract` and `mi replace0`.

Also see

[MI] **intro** — Introduction to mi

[MI] **mi extract** — Extract original or imputed data from mi data

Title

> **mi reset** — Reset imputed or passive variables

Syntax

> mi reset *varlist* [= *exp* [*if*] [, *options*]]

options	description
Main	
m(*numlist*)	m to reset; default all
noupdate	see [MI] **noupdate option**

Menu

Statistics > Multiple imputation

Description

mi reset resets the imputed or passive variables specified. Values are reset to the values in $m = 0$, which are typically missing, but if you specify = *exp*, they are reset to the value specified.

Options

    ⌐ Main ⌐

m(*numlist*) specifies the values of m that are to be reset; the default is to update all values of m. If M were equal to 3, the default would be equivalent to specifying m(1/3) or m(1 2 3). If you wished to update the specified variable(s) in just $m = 2$, you could specify m(2).

noupdate in some cases suppresses the automatic mi update this command might perform; see [MI] **noupdate option**.

Remarks

Remarks are presented under the following headings:

> Using mi reset
> Technical notes and relation to mi update

Using mi reset

Resetting an imputed or passive variable means setting its values in $m > 0$ equal to the values recorded in $m = 0$. For instance, if variable inc were imputed, typing

```
. mi reset inc
(15 values reset)
```

would reset its incomplete values back to missing in all m. In the sample output shown, we happen to have $M = 5$ and reset back to missing the three previously imputed values in each imputation.

It is rare that you would want to reset an imputed variable, but one can imagine cases. Your coworker Joe sent you the data and just buzzed you on the telephone. "There is one value wrong in the data I sent you," he says. "There is an imputed value for inc that is 15,000, which is obviously absurd. Just reset it back to missing until I find out what happened." So you type

```
. mi reset inc if inc==15000
(1 value reset)
```

Later Joe calls back. "It is a long and very stupid story," he begins, and you can hear him settling into his chair to tell it. As you finish your second cup of coffee, he is wrapping up. "So the value of inc for pid 1433 should be 0.725." You type

```
. mi reset inc = .725 if pid=1433
(1 value reset)
```

It is common to need to reset passive variables if imputed values change. For instance, you have variables age and lnage in your data. You imputed lnage; age is passive. You recently updated the imputed values for lnage. One way to update the values for age would be to type

```
. mi passive: replace age = exp(lnage)
m=0:
m=1:
(10 real changes made)
m=2:
(10 real changes made)
m=3:
(8 real changes made)
```

Alternatively, you could type

```
. mi reset age = exp(lnage)
(28 values reset)
```

Technical notes and relation to mi update

mi reset, used with an imputed variable, changes only the values for which the variable contains hard missing (.) in $m = 0$. The other values are, by definition, already equal to their $m = 0$ values.

mi reset, used with a passive variable, changes only the values in incomplete observations, observations in which any imputed variable contains hard missing. The other values of the passive variable are, by definition, already equal to their $m = 0$ values.

mi update can be used to ensure that values that are supposed to be equal to their $m = 0$ values in fact are equal to them; see [MI] **mi update**.

Also see

[MI] **intro** — Introduction to mi

[MI] **mi update** — Ensure that mi data are consistent

Title

> **mi reshape** — Reshape mi data

Syntax

(The words long and wide in what follows have nothing to do with mi styles mlong, flong, flongsep, and wide; they have to do with reshape's concepts.)

    mi reshape wide *stubnames*, i(*varlist*) j(*varname*) [*options*]

    mi reshape long *stubnames*, i(*varlist*) j(*varname*) [*options*]

long

i	j	stub
1	1	4.1
1	2	4.5
2	1	3.3
2	2	3.0

reshape ⟷

wide

i	stub1	stub2
1	4.1	4.5
2	3.3	3.0

To go from long to wide:

    mi reshape wide *stub*, i(*i*) j(*j*)

where *j* is an existing variable.

To go from wide to long:

    mi reshape long *stub*, i(*i*) j(*j*)

where *j* is a new variable.

options	description
i(*varlist*)	*i* variable(s)
j(*varname* [*values*])	long→wide: *j*, existing variable wide→long: *j*, new variable optionally specify values to subset *j*
<u>s</u>tring	*j* is string variable (default is numeric)

where *values* is   #[-#] [...]   if *j* is numeric (the default)
                    "*string*" ["*string*" ...]   if *j* is string

and where *stubnames* are variable names (long→wide), or stubs of variable names (wide→long), and either way, may contain @, denoting where *j* appears or is to appear in the name.

In the example above, when we wrote "mi reshape wide *stub*", we could have written "mi reshape wide *stub*@" because *j* by default is used as a suffix. Had we written *stu*@*b*, then the wide variables would have been named *stu*1*b* and *stu*2*b*.

Menu

Statistics > Multiple imputation

Description

mi reshape is Stata's reshape for mi data; see [D] **reshape**.

Options

noupdate in some cases suppresses the automatic mi update this command might perform; see [MI] **noupdate option**.

See [D] **reshape** for descriptions of the other options.

Remarks

The reshape command you specify is carried out on the $m = 0$ data, and then the result is duplicated in $m = 1$, $m = 2$, ..., $m = M$.

Also see

[MI] **intro** — Introduction to mi

[D] **reshape** — Convert data from wide to long form and vice versa

[MI] **mi replace0** — Replace original data

Title

> **mi select** — Programmer's alternative to mi extract

Syntax

   mi select init [, fast]

   mi select #

where $0 \leq \# \leq M$, and where typical usage is

            quietly mi query
            local M = r(M)
            preserve
            mi select init
            local priorcmd "`r(priorcmd)'"
            forvalues m=1(1)`M' {
                    mi select `m'
                    ...
                    `priorcmd'
            }
            restore

Description

   mi select is a programmer's command. It is a faster, more dangerous version of mi extract; see [MI] **mi extract**.

   Before using mi select, the mi data must be preserved; see [P] **preserve**.

   mi select init initializes mi select. mi select returns macro r(priorcmd), which you are to issue as a command between each mi select # call. r(priorcmd) is not required to be issued before the first call to mi select #, although you may issue it if that is convenient.

   mi select # replaces the data in memory with a copy of the data for $m = \#$. The data are not mi set. Changes to the selected data will not be posted back to the underlying mi data. mi select # calls can be made in any order, and the same m may be selected repeatedly.

Option

   fast, specified with mi select init, specifies that the data delivered by mi select # commands not be changed except for sort order. Then mi select can operate more quickly. fast is allowed with all styles but currently affects the performance with the wide style only.

   If fast is not specified, the data delivered by mi select # may be modified freely before the next mi select # call. However, the data may not be dropped. mi select uses characteristics stored in _dta[] to know its state.

Remarks

The data delivered by mi select # differ from those delivered by mi extract in that there may be extra variables in the dataset.

One of the extra variables, _mi_id, is a unique observation identifier that you can use. If you want to post changes made in the selected data back to the mi data, you can write a file containing mi_id and the updated variables and then use _mi_id to match that to the mi data after your final restore.

In the case of wide data, the mi data have no _mi_id variable. _mi_id in the selected data is reflected in the current order of the mi data.

Saved results

mi select init returns the following in r():

Macros
    r(priorcmd) command to be issued prior to calling mi select #; this command will be either restore, preserve or nothing.

Also see

[MI] **intro** — Introduction to mi

[MI] **mi extract** — Extract original or imputed data from mi data

[MI] **technical** — Details for programmers

Title

> **mi set** — Declare multiple-imputation data

Syntax

    mi set *style*

        where *style* is   <u>w</u>ide
                            <u>ml</u>ong
                            <u>fl</u>ong
                            <u>fl</u>ongsep *name*

    mi <u>reg</u>ister { <u>imp</u>uted | <u>pas</u>sive | <u>reg</u>ular } *varlist*

    mi <u>unreg</u>ister *varlist*

    mi set M { = | += | -= } #

    mi set m -= (*numlist*)

    mi unset [, asis]

Menu

Statistics > Multiple imputation

Description

    mi set is used to set a regular Stata dataset to be an mi dataset. mi set is also used to modify the attributes of an already set dataset. An mi set dataset has the following attributes:

- The data are recorded in a *style*: wide, mlong, flong, or flongsep.
- Variables are registered as imputed, passive, or regular, or they are left unregistered.
- In addition to $m = 0$, the data with missing values, the data include $M \geq 0$ imputations of the imputed variables.

    mi set *style* begins the setting process by setting the desired style. mi set *style* sets all variables as unregistered and sets $M = 0$.

    mi register registers variables as imputed, passive, or regular. Variables can be registered one at a time or in groups and can be registered and reregistered.

    mi unregister unregisters registered variables, which is useful if you make a mistake. Exercise caution. Unregistering an imputed or passive variable can cause loss of the filled-in missing values in $m > 0$ if your data are recorded in the wide or mlong styles. In such cases, just mi register the variable correctly without mi unregistering it first.

mi set M modifies M, the total number of imputations. M may be increased or decreased. M may be set before or after imputed variables are registered.

mi set m drops selected imputations from the data.

mi unset is a rarely used command to unset the data. Better alternatives include mi extract and mi export (see [MI] **mi extract** and [MI] **mi export**, respectively).

Remarks

Data must be mi set before they can be used with the other mi commands. There are two ways data can become mi set: direct use of mi set *style* or use of mi import (see [MI] **mi import**).

The mi register, mi set M, and mi set m commands are for use with already set data and are useful even with imported data.

Remarks are presented under the following headings:

    mi set style
    mi register and mi unregister
    mi set M and mi set m
    mi unset

mi set style

mi set *style* begins the setting process. mi set *style* has the following forms:

    mi set wide
    mi set mlong
    mi set flong
    mi set flongsep *name*

It does not matter which style you choose because you can always use mi convert (see [MI] **mi convert**) to change the style later. We typically choose wide to begin.

If your data are large, you may have to use flongsep. mi set flongsep requires you to specify a name for the flongsep dataset collection. See *Advice for using flongsep* in [MI] **styles**.

If you intend to have super-varying variables, you need to choose either flong or flongsep, or mi convert to flong or flongsep style later.

The current style of the data is shown by the mi query and mi describe commands; see [MI] **mi describe**.

mi register and mi unregister

mi register has three forms:

    mi register imputed *varlist*
    mi register passive *varlist*
    mi register regular *varlist*

See [MI] **Glossary** for a definition of imputed, passive, and regular variables.

You are required to register imputed variables. If you intend to use mi impute (see [MI] **mi impute**) to impute missing values, you must still register the variables first.

Concerning passive variables, we recommend that you register them, and if your data are style wide, you are required to register them. If you create passive variables by using mi passive (see [MI] **mi passive**), that command automatically registers them for you.

Whether you register regular variables is up to you. Registering them is safer in all styles except wide, where it does not matter. We say registering is safer because regular variables should not vary across m, and in the long styles, you can unintentionally create variables that vary. If variables are registered, mi will detect and fix mistakes for you.

Super-varying variables—see [MI] **Glossary**—rarely occur, but if you have them, be aware that they can be stored only in flong and flongsep data and that they never should be registered.

The registration status of variables is listed by mi describe (see [MI] **mi describe**).

Use mi unregister if you accidentally register a variable incorrectly, with one exception: if you mistakenly register a variable as imputed but intended to register it as passive, or vice versa, use mi register directly to reregister the variable. The mere act of unregistering a passive or imputed variable can cause values in $m > 0$ to be replaced with those from $m = 0$ if the data are wide or mlong.

That exception aside, you first mi unregister variables before reregistering them.

mi set M and mi set m

mi set M is seldom used, and mi set m is sometimes used.

mi set M sets M, the total number of imputations. The syntax is

    mi set M = #
    mi set M += #
    mi set M -= #

mi set M = # sets $M = \#$. Imputations are added or deleted as necessary. If imputations are added, the new imputations obtain their values of imputed and passive variables from $m = 0$, which means that the missing values are not yet replaced in the new imputations. It is not necessary to increase M if you intend to use mi impute to impute values; see [MI] **mi impute**.

mi set M += # increments M by #.

mi set M -= # decrements M by #.

mi set m -= (*numlist*) deletes the specified imputations. For instance, if you had $M = 5$ imputations and wanted to delete imputation 2, leaving you with $M = 4$, you would type mi set m -= (2).

mi unset

If you wish to unset your data, your best choices are mi extract and mi export; see [MI] **mi extract** and [MI] **mi export**. The mi extract 0 command replaces the data in memory with the data from $m = 0$, unset. The mi export command replaces the data in memory with unset data in a form that can be sent to a non–Stata user.

mi unset is included for completeness, and if it has any use at all, it would be by programmers.

Also see

[MI] **intro** — Introduction to mi

[MI] **mi convert** — Change style of mi data

[MI] **mi describe** — Describe mi data

[MI] **mi extract** — Extract original or imputed data from mi data

[MI] **mi export** — Export mi data

[MI] **mi import** — Import data into mi

[MI] **styles** — Dataset styles

Title

> **mi stsplit** — Stsplit and stjoin mi data

Syntax

To split at designated times

    mi stsplit *newvar* [*if*] , { at(*numlist*) | <u>ev</u>ery(*#*) } [*options*]

options	description
Main	
* at(*numlist*)	split at specified analysis times
* <u>ev</u>ery(*numlist*)	split when analysis time is a multiple of *#*
<u>af</u>ter(*spec*)	use time since *spec* instead of analysis time for at() or every()
trim	exclude observations outside of range
<u>noup</u>date	see [MI] **noupdate option**
† <u>nop</u>reserve	programmer's option

* at() or every() is required.

† nopreserve is not included in the dialog box.

To split at failure times

    mi stsplit [*if*] , at(<u>f</u>ailures) [*options*]

options	description
Main	
* at(<u>f</u>ailures)	split at times of observed failures
<u>s</u>trata(*varlist*)	perform splitting by failures within stratum, strata defined by *varlist*
<u>r</u>iskset(*newvar*)	create risk-set ID variable
<u>noup</u>date	see [MI] **noupdate option**
† <u>nop</u>reserve	programmer's option

* at() is required.

† nopreserve is not included in the dialog box.

216   mi stsplit — Stsplit and stjoin mi data

To join episodes

    mi stjoin [, *options*]

options	description
Main	
<u>c</u>ensored(*numlist*)	values of failure that indicate no event
<u>noup</u>date	see [MI] **noupdate option**

Menu

Statistics > Multiple imputation

Description

    mi stsplit and mi stjoin are stsplit and stjoin for mi data; see [ST] **stsplit**. Except for the addition of the noupdate option, the syntax is identical. Except for generalization across m, the results are identical.

    Your mi data must be stset to use these commands. If your data are not already stset, use mi stset rather than the standard stset; see [MI] **mi XXXset**.

Options

    noupdate in some cases suppresses the automatic mi update this command might perform; see [MI] **noupdate option**.

    See [ST] **stsplit** for documentation on the remaining options.

Remarks

    One should never use any heavyweight data-management commands with mi data. Heavyweight commands are commands that make sweeping changes to the data rather than simply deleting some observations, adding or dropping some variables, or changing some values of existing variables. stsplit and stjoin are examples of heavyweight commands (see [ST] **stsplit**).

Also see

    [MI] **intro** — Introduction to mi

    [ST] **stsplit** — Split and join time-span records

    [MI] **mi XXXset** — Declare mi data to be svy, st, ts, xt, etc.

Title

> **mi update** — Ensure that mi data are consistent

Syntax

    mi update

Menu

Statistics > Multiple imputation

Description

mi update verifies that mi data are consistent. If the data are not consistent, mi update reports the inconsistencies and makes the necessary changes to make the data consistent.

mi update can change the sort order of the data.

Remarks

Remarks are presented under the following headings:

> *Purpose of mi update*
> *What mi update does*
> *mi update is run automatically*

Purpose of mi update

mi update allows you to

- change the values of existing variables, whether imputed, passive, regular, or unregistered;
- add or remove missing values from imputed variables (or from any variables);
- drop variables;
- create new variables;
- drop observations; and
- duplicate observations (but not add observations in other ways).

You can make any or all of the above changes and then type

    . mi update

and mi update will handle making whatever additional changes are required to keep the data consistent. For instance,

    . drop if sex==1
    (75 observations deleted)
    . mi update
    (375 m>0 obs. dropped due to dropped obs. in m=0)

In this example, we happen to have five imputations and are working with flongsep data. We dropped 75 observations in $m = 0$, and that still left $5 \times 75 = 375$ observations to be dropped in $m > 0$.

The messages `mi update` produces vary according to the style of the data because the changes required to make the data consistent are determined by the style. Had we been working with flong data, we would have seen

```
. drop if sex==1
(450 observations deleted)
. mi update
(system variable _mi_id updated due to change in number of obs.)
```

With flong data in memory, when we dropped `if sex==1`, we dropped all $75 + 5 \times 75 = 450$ observations, so no more observations needed to be dropped; but here `mi update` needed to update one of its system variables because of the change we made.

Had we been working with mlong data, we might have seen

```
. drop if sex==1
(90 observations deleted)
. mi update
(system variable _mi_id updated due to change in number of obs.)
```

The story here is very much like the story in the flong case. In mlong data, dropping `if sex==1` drops the 75 observations in $m = 0$ and also drops the incomplete observations among the 75 in $m = 1$, $m = 2$, ..., $m = 5$. In this example, there are 3 such observations, so a total of $75 + 5 \times 3 = 90$ were dropped, and because of the change, `mi update` needed to update its system variable.

Had we been using wide data, we would have seen

```
. drop if sex==1
(75 observations deleted)
. mi update
```

`mi update`'s silence indicates that `mi update` did nothing because after dropping observations in wide data, nothing more needs to be done. We could have skipped typing `mi update` here, but do not think that way because changing values, dropping variables, creating new variables, dropping observations, or creating new observations can have unanticipated consequences.

For instance, in our data is variable `farmincome`, and it seems obvious that `farmincome` should be 0 if the person does not have a farm, so we type

```
. replace farmincome = 0 if !farm
(15 real changes made)
```

After changing values, you should type `mi update` even if you do not suspect that it is necessary. Here is what happens when we do that with these data:

```
. mi update
(12 m=0 obs. now marked as complete)
```

Typing `mi update` was indeed necessary! We forgot that the `farmincome` variable was imputed, and it turns out that the variable contained missing in 12 nonfarm observations; `mi` needed to deal with that.

Running `mi update` is so important that `mi` itself is constantly running it just in case you forget. For instance, let's "forget" to type `mi update` and then convert our data to wide:

```
. replace farmincome = 0 if !farm
(15 real changes made)
. mi convert wide, clear
(12 m=0 obs. now marked as complete)
```

The parenthetical message was produced because `mi convert` ran `mi update` for us. For more information on this, see [MI] **noupdate option**.

What mi update does

- `mi update` checks whether you have changed N, the number of observations in $m = 0$, and resets N if necessary.
- `mi update` checks whether you have changed M, the number of imputations, and adjusts the data if necessary.
- `mi update` checks whether you have added, dropped, registered, or unregistered any variables and takes the appropriate action.
- `mi update` checks whether you have added or deleted any observations. If you have, it then checks whether you carried out the operation consistently for $m = 0, m = 1, \ldots, m = M$. If you have not carried it out consistently, `mi update` carries it out consistently for you.
- In the mlong, flong, and flongsep styles, `mi update` checks system variable _mi_id, which links observations across m, and reconstructs the variable if necessary.
- `mi update` checks that the system variable _mi_miss, which marks the incomplete observations, is correct and, if not, updates it and makes any other changes required by the change.
- `mi update` verifies that the values recorded in imputed variables in $m > 0$ are equal to the values in $m = 0$ when they are nonmissing and updates any that differ.
- `mi update` verifies that the values recorded in passive variables in $m > 0$ are equal to the values recorded in $m = 0$'s complete observations and updates any that differ.
- `mi update` verifies that the values recorded in regular variables in $m > 0$ equal the values in $m = 0$ and updates any that differ.
- `mi update` adds any new variables in $m = 0$ to $m > 0$.
- `mi update` drops any variables from $m > 0$ that do not appear in $m = 0$.

mi update is run automatically

As we mentioned before, running `mi update` is so important that many `mi` commands simply run it as a matter of course. This is discussed in [MI] **noupdate option**. In a nutshell, the `mi` commands that do this have a `noupdate` option, so you can identify them, and you can specify the option to skip running the update and so speed execution, but only with the adrenaline rush caused by a small amount of danger.

Whether you specify that option or not, we advise you to run `mi update` periodically, and to always run `mi update` after dropping or adding variables or observations, or changing values.

Also see

[MI] **intro** — Introduction to mi

[MI] **noupdate option** — The noupdate option

Title

> **mi varying** — Identify variables that vary across imputations

Syntax

mi varying [*varlist*] [, noupdate]

mi varying, unregistered [noupdate]

Menu

Statistics > Multiple imputation

Description

mi varying lists the names of variables that are unexpectedly varying and super varying; see [MI] **Glossary** for a definition of varying and super-varying variables.

Options

unregistered specifies that the listing be made only for unregistered variables. Specifying this option saves time, especially when the data are flongsep.

noupdate in some cases suppresses the automatic mi update this command might perform; see [MI] **noupdate option**.

Remarks

A variable is said to be varying if it varies over m in the complete observations. A variable is said to be super varying if it varies over m in the incomplete observations.

Remarks are presented under the following headings:

> *Detecting problems*
> *Fixing problems*

Detecting problems

mi varying looks for five potential problems:

1. *Imputed nonvarying.* Variables that are registered as imputed and are nonvarying either

    a. do not have their missing values in $m > 0$ filled in yet, in which case you should use mi impute (see [MI] **mi impute**) to impute them, or

    b. have no missing values in $m = 0$, in which case you should mi unregister the variables and perhaps use mi register to register the variables as regular (see [MI] **mi set**).

220

2. *Passive nonvarying.* Variables that are registered as passive and are nonvarying either

   a. have missing values in the incomplete observations in $m > 0$, in which case after you have filled in the missing values of your imputed variables, you should use mi passive (see [MI] **mi passive**) to update the values of these variables, or

   b. have no missing values in $m = 0$, in which case you should mi unregister the variables and perhaps use mi register to register the variables as regular (see [MI] **mi set**).

3. *Unregistered varying.*

   a. It is most likely that such variables should be registered as imputed or as passive.

   b. If the variables are varying but should not be, use mi register to register them as regular. That will fix the problem; values from $m = 0$ will be copied to $m > 0$.

   c. It is possible that this is just like case 5 below and it just randomly turned out that the only observations in which variation occurred were the incomplete observations. In that case, leave the variable unregistered.

4. *Unregistered super/varying.* These are variables that are super varying but would have been categorized as varying if they were registered as imputed. This is to say that while they have varying values in the complete observations *as complete is defined this instant*—which is based on the variables currently registered as imputed—these variables merely vary in observations for which they themselves contain missing in $m = 0$, and thus they could be registered as imputed without loss of information. Such variables should be registered as imputed.

5. *Unregistered super varying.* These variables really do super vary and could not be registered as imputed without loss of information. These variables either contain true errors or they are passive variables that are functions of groups of observations. Fix the errors by registering the variables as regular and leave unregistered those intended to be super varying. If you intentionally have super-varying variables in your data, remember never to convert to the wide or mlong styles. Super-varying variables can appear only in the flong and flongsep styles.

mi varying output looks like this:

```
                 Possible problem    Variable names
         imputed nonvarying:         (none)
         passive nonvarying:         (none)
        unregistered varying:        (none)
   *unregistered super/varying:      (none)
    unregistered super varying:      (none)
```

  * super/varying means super varying but would be varying if registered as imputed; variables vary only where equal to soft missing in m=0.

If there are possible problems, variable names are listed in the table.

Super-varying variables can arise only in flong and flongsep data, so the last two categories are omitted when mi varying is run on wide or mlong data. If there are no imputed variables, or no passive variables, or no unregistered variables, the corresponding categories are omitted from the table.

Fixing problems

If mi varying detects problems, register all imputed variables before registering passive variables. Rerun mi varying as you register new imputed variables. Registering new variables as imputed can change which observations are classified as complete and incomplete, and that classification in turn can change the categories to which the other variables are assigned. After registering a variable as imputed, another variable previously listed as super varying might now be merely varying.

Saved results

mi varying saves the following in r():

Macros
r(ivars)	nonvarying imputed variables
r(pvars)	nonvarying passive variables
r(uvars_v)	varying unregistered variables
r(u_uvars_s_v)	(super) varying unregistered variables
r(uvars_s_s)	super-varying unregistered variables

Also see

[MI] **intro** — Introduction to mi

[MI] **mi misstable** — Tabulate pattern of missing values

Title

> **mi xeq** — Execute command(s) on individual imputations

Syntax

mi xeq [*numlist*] : *command* [; *command* [; ...]]

Description

mi xeq: *XXX* executes *XXX* on $m = 0$, $m = 1$, ..., $m = M$.

mi xeq *numlist*: *XXX* executes *XXX* on $m = numlist$.

XXX can be any single command or it can be multiple commands separated by a semicolon. If specifying multiple commands, the delimiter must not be set to semicolon; see [P] **#delimit**.

Remarks

Remarks are presented under the following headings:

> Using mi xeq with reporting commands
> Using mi xeq with data-modification commands
> Using mi xeq with data-modification commands on flongsep data

Using mi xeq with reporting commands

By reporting commands, we mean any general Stata command that reports results but leaves the data unchanged. summarize (see [R] **summarize**) is an example. mi xeq is especially useful with such commands. If you wanted to see the summary statistics for variables outcome and age among the females in your mi data, you could type

```
. mi xeq: summarize outcome age if sex=="female"
m=0 data:
-> summarize outcome age if sex=="female"
    (output omitted)
m=1 data:
-> summarize outcome age if sex=="female"
    (output omitted)
m=2 data:
-> summarize outcome age if sex=="female"
    (output omitted)
```

$M = 2$ in the data above.

If you wanted to see a particular regression run on the $m = 2$ data, you could type

```
. mi xeq 2: regress outcome age bp
m=2 data:
-> regress outcome age bp
    (output omitted)
```

In both cases, once the command executes, the entire mi dataset is brought back into memory.

Using mi xeq with data-modification commands

You can use data-modification commands with `mi xeq` but doing that is not especially useful unless you are using flongsep data.

If variable `lnage` were registered as passive and you wanted to update its values, you could type

```
. mi xeq: replace lnage = ln(age)
```
(*output omitted*)

That would work regardless of style, although it is just as easy to update the variable using `mi passive` (see [MI] **mi passive**):

```
. mi passive: replace lnage = ln(age)
```
(*output omitted*)

If what you are doing depends on the sort order of the data, include the `sort` command among the commands to be executed; do not assume that the individual datasets will be sorted the way the data in memory are sorted. For instance, if you have passive variable `totalx`, do not type

```
. sort id time
. mi xeq: by id: replace totalx = sum(x)
```

That will not work. Instead, type

```
. mi xeq: sort id time;  by id: replace totalx = sum(x)
m=0 data:
-> sort id time
-> by id: replace total x = sum(x)
(8 changes made)
m=1 data:
-> sort id time
-> by id: replace total x = sum(x)
(8 changes made)
m=2 data:
-> sort id time
-> by id: replace total x = sum(x)
(8 changes made)
```

Again we note that it would be just as easy to update this variable with `mi passive`:

```
. mi passive: by id (time): replace totalx = sum(x)
m=0:
(8 changes made)
m=1:
(8 changes made)
m=2:
(8 changes made)
```

With the wide, mlong, and flong styles, there is always another way to proceed, and often the other way is easier.

Using mi xeq with data-modification commands on flongsep data

With flongsep data, `mi xeq` is especially useful. Consider the case where you want to add new variable `lnage = ln(age)` to your data, and `age` is just a regular or unregistered variable. With flong, mlong, or wide data, you would just type

```
. generate lnage = ln(age)
```

and be done with it.

With flongsep data, you have multiple datasets to update. Of course, you could mi convert (see [MI] **mi convert**) your data to one of the other styles, but we will assume that if you had sufficient memory to do that, you would have done that long ago and so would not be using flongsep data.

The easy way to create lnage with flongsep data is by typing

```
. mi xeq: gen lnage = ln(age)
```
(*output omitted*)

You could use the mi xeq approach with any of the styles, but with flong, mlong, or wide data, it is not necessary. With flongsep, it is.

Saved results

mi xeq saves in r() whatever the last command run on the last imputation or specified imputation returns. For instance,

```
. mi xeq: tabulate g ;  summarize x
```

returns the saved results for summarize x run on $m = M$.

```
. mi xeq 1 2: tabulate g ;  summarize x
```

returns the saved results for summarize x run on $m = 2$.

```
. mi xeq 0: summarize x
```

returns the saved results for summarize x run on $m = 0$.

Also see

[MI] **intro** — Introduction to mi

[MI] **mi passive** — Generate/replace and register passive variables

Title

> **mi XXXset** — Declare mi data to be svy, st, ts, xt, etc.

Syntax

    mi fvset ...   see [R] **fvset**

    mi svyset ...   see [SVY] **svyset**

    mi stset ...   see [ST] **stset**
    mi streset ...
    mi st ...

    mi tsset ...   see [TS] **tsset**
    mi xtset ...   see [XT] **xtset**

Description

Using some features of Stata requires setting your data. The commands listed above allow you to do that with `mi` data. The `mi` variants have the same syntax and work the same way as the original commands.

Remarks

If you have set your data with any of the above commands before you `mi set` them, there is no problem; the settings were automatically imported. Once you `mi set` your data, however, you will discover that Stata's other set commands no longer work. For instance, here is the result of typing `stset` on an `mi set` dataset:

```
. stset ...
no; data are mi set
    Use mi stset to set or query these data; mi stset has the same
    syntax as stset.
r(119);
```

The solution is to use `mi stset`:

```
. mi stset ...
```
(*usual output appears*)

After `mi` setting your data, put `mi` in front of Stata's other set commands.

Also, you might sometimes see an error like the one above when you give a command that depends on the data being set by one of Stata's other set commands. In general, it is odd that you would be running such a command directly on `mi` data because what you will get will depend on the `mi` style of data. Perhaps, however, you are using `mi` wide data, where the structure of the data more or less corresponds to the structure of non-mi data, or perhaps you have smartly specified the appropriate if statement to account for the `mi` style of data you are using. In any case, the result might be

```
        . some_other_command
        no; data are mi set
            Use mi XXXset to set or query these data; mi XXXset has the same
            syntax as XXXset.
        r(119);
```

Substitute one of the set commands listed above for XXXset, and then understand what just happened. You correctly used `mi XXXset` to set your data, you thought your data were set, and yet, when you tried to use a command that depended on the data being XXXset, you received this error.

If this happens to you, the solution is to use `mi extract` (see [MI] **mi extract**) to obtain the data on which you want to run the command—which is probably $m = 0$, so you would type `mi extract 0`—and then run the command.

Also see

[MI] **intro** — Introduction to mi

Title

noupdate option — The noupdate option

Syntax

mi ... [, ... <u>noup</u>date ...]

Description

Many mi commands allow the noupdate option. This entry describes the purpose of that option.

Option

noupdate specifies that the mi command in question need not perform an mi update because you are certain that there are no inconsistencies that need fixing; see [MI] **mi update**. noupdate is taken as a suggestion; mi update will still be performed if the command sees evidence that it needs to be. Not specifying the option does not mean that an mi update will be performed.

Remarks

Some mi commands perform modifications to the data, and those modifications will go very poorly—even to the point of corrupting your data—if certain assumptions about your data are not true. Usually, those assumptions are true, but to be safe, the commands check the assumptions. They do this by calling mi update; see [MI] **mi update**. mi update checks the assumptions and, if they are not true, corrects the data so that the assumptions are true. mi update always reports the data corrections it makes.

All of this reflects an abundance of caution, with the result that some commands spend more time running mi update than they spend performing their intended task.

Commands that use mi update to verify assumptions have a noupdate option. When you specify that option, the command skips checking the assumptions, which is to say, skips calling mi update. More correctly, the command skips calling mi update if the command sees no obvious evidence that mi update needs to be called.

You can make commands run faster by specifying noupdate. Should you? Unless you are noticing poor performance, we would say no. It is, however, absolutely safe to specify noupdate if the only commands executed since the last mi update are mi commands. The following would be perfectly safe:

```
. mi update
. mi passive, noupdate: gen agesq = age*age
. mi rename age age_at_admission, noupdate
. mi ...
```

The following would be safe, too:

```
. mi update
. mi passive, noupdate: gen agesq = age*age
. summarize agesq
. mi rename age age_at_admission, noupdate
. mi ...
```

It would be safe because `summarize` is a reporting command that does not change the data.

The problem `mi` has is that it is not in control of your session and data. Between `mi` commands, `mi` does not know what you have done to the data. The following would not be recommended and has the potential to go very poorly:

```
. mi update
. mi passive, noupdate: gen agesq = age*age
. drop if female
. drop agesq
. mi ..., noupdate              // do not do this
```

By the rules for using `mi`, you should perform an `mi update` yourself after a `drop` command, or any other command that changes the data, but it usually does not matter whether you follow that rule because `mi` will check eventually, when it matters. That is, `mi` will check if you do not specify the `noupdate` option.

The `noupdate` option is recommended for use by programmers in programs that code a sequence of `mi` commands.

Also see

[MI] **intro** — Introduction to mi

[MI] **mi update** — Ensure that mi data are consistent

Title

styles — Dataset styles

Syntax

There are four dataset styles available for storing mi data:

    <u>w</u>ide

    <u>ml</u>ong

    <u>fl</u>ong

    <u>flongs</u>ep

Description

The purpose of this entry is to familiarize you with the four styles in which mi data can be stored.

Remarks

Remarks are presented under the following headings:

    The four styles
       Style wide
       Style flong
       Style mlong
       Style flongsep
       How we constructed this example
    Using mi system variables
    Advice for using flongsep

The four styles

We have highly artificial data, which we will first describe verbally and then show to you in each of the styles. The original data have two observations on two variables:

a	b
1	2
4	.

Variable b has a missing value. We have two imputed values for b, namely, 4.5 and 5.5. There will also be a third variable, c, in our dataset, where $c = a + b$.

Thus, in the jargon of `mi`, we have $M = 2$ imputations, and the datasets $m = 0$, $m = 1$, and $m = 2$ are

```
m=0:    a       b       c
        1       2       3
        4       .       .


m=1:    a       b       c
        1       2       3
        4       4.5     8.5


m=2:    a       b       c
        1       2       3
        4       5.5     9.5
```

Continuing with jargon, a is a regular variable, b is an imputed variable, and c is a passive variable.

Style wide

The above data have been stored in `miproto.dta` in the wide style.

```
. use http://www.stata-press.com/data/r11/miproto
(mi prototype)
. list

     a    b    c    _1_b    _2_b    _1_c    _2_c    _mi_miss
1.   1    2    3      2       2       3       3        0
2.   4    .    .    4.5     5.5     8.5     9.5        1
```

There is no significance to the order in which the variables appear.

On the left, under variables a, b, and c, you can see the original data.

The imputed values for b appear under the variables named _1_b and _2_b; $m = 1$ appears under _1_b, and $m = 2$ appears under _2_b. Note that in the first observation, the observed value of b is simply repeated in _1_b and _2_b. In the second observation, however, _1_b and _2_b show the replacement values for the missing value of b.

The passive values for c appear under the variables named _1_c and _2_c in the same way that the imputed values appeared under the variables named _1_b and _2_b.

Finally, one extra variable appears: _mi_miss. This is an example of an `mi` system variable. You are never to change `mi` system variables; they take care of themselves. The wide style has only one system variable. _mi_miss contains 0 for complete observations and 1 for incomplete observations.

Style flong

Let's convert this dataset to style flong:

```
. mi convert flong, clear
. list, separator(2)
```

	a	b	c	_mi_miss	_mi_m	_mi_id
1.	1	2	3	0	0	1
2.	4	.	.	1	0	2
3.	1	2	3	.	1	1
4.	4	4.5	8.5	.	1	2
5.	1	2	3	.	2	1
6.	4	5.5	9.5	.	2	2

We listed these data with a separator line after every two rows so that they would be easier to understand. Ignore the mi system variables and focus on variables a, b, and c. Observations 1 and 2 contain $m = 0$; observations 3 and 4 contain $m = 1$; observations 5 and 6 contain $m = 2$.

We will now explain the system variables, but you do not need to remember this.

1. We again see _mi_miss, just as we did in the wide style. It marks the incomplete observations in $m = 0$. It contains missing in $m > 0$.

2. _mi_m records m. The first two observations are $m = 0$; the next two, $m = 1$; and the last two, $m = 2$.

3. _mi_id records an arbitrarily coded observation-identification variable. It is 1 and 2 in $m = 0$, and then repeats in $m = 1$ and $m = 2$. Observations _mi_id = 1 correspond to each other for all m. The same applies to _mi_id = 2.

   Warning: Do not use _mi_id as your own ID variable. You might look one time, see that a particular observation has _mi_id = 8, and look a little later, and see that the observation has changed from _mi_id = 8 to _mi_id = 5. _mi_id belongs to mi. If you want your own ID variable, make your own. All that is true of _mi_id is that, at any instant, it uniquely identifies, and ties together, the observations.

There is no significance to the order of the variables or, for that matter, to the order of the observations.

Style mlong

Let's convert this dataset to the mlong style:

```
. mi convert mlong, clear
. list
```

	a	b	c	_mi_miss	_mi_m	_mi_id
1.	1	2	3	0	0	1
2.	4	.	.	1	0	2
3.	4	4.5	8.5	.	1	2
4.	4	5.5	9.5	.	2	2

This listing will be easier to read if we add some carefully chosen blank lines:

	a	b	c	_mi_miss	_mi_m	_mi_id
1.	1	2	3	0	0	1
2.	4	.	.	1	0	2
3.	4	4.5	8.5	.	1	2
4.	4	5.5	9.5	.	2	2

The mlong style is just like flong except that the complete observations—observations for which $\_mi\_miss = 0$ in $m = 0$—are omitted in $m > 0$.

Observations 1 and 2 are the original, $m = 0$ data.

Observation 3 is the $m = 1$ replacement observation for observation 2.

Observation 4 is the $m = 2$ replacement observation for observation 2.

Style flongsep

Let's look at these data in the flongsep style:

```
. mi convert flongsep example, clear
(files example.dta _1_example.dta _2_example.dta created)
. list
```

	a	b	c	_mi_miss	_mi_id
1.	1	2	3	0	1
2.	4	.	.	1	2

The flongsep style stores $m = 0$, $m = 1$, and $m = 2$ in separate files. When we converted to the flongsep style, we had to specify a name for these files, and we chose example. This resulted in $m = 0$ being stored in example.dta, $m = 1$ being stored in _1_example.dta, and $m = 2$ being stored in _2_example.dta.

In the listing above, we see the original, $m = 0$ data.

After conversion, $m = 0$ (example.dta) was left in memory. When working with flongsep data, you always work with $m = 0$ in memory. Nothing can stop us, however, from taking a brief peek:

```
. save example, replace
file example.dta saved
. use _1_example, clear
(mi prototype)
. list
```

	a	b	c	_mi_id
1.	1	2	3	1
2.	4	4.5	8.5	2

There are the data for $m = 1$. As previously, system variable $\_mi\_id$ ties together observations. In the $m = 1$ data, however, $\_mi\_miss$ is not repeated.

Let's now look at _2_example.dta:

```
. use _2_example, clear
(mi prototype)
. list
```

	a	b	c	_mi_id
1.	1	2	3	1
2.	4	5.5	9.5	2

And there are the data for $m = 2$.

We have an aside, but an important one. Review the commands we just gave, stripped of their output:

```
. mi convert flongsep example, clear
. list
. save example, replace
. use _1_example, clear
. list
. use _2_example, clear
. list
```

What we want you to notice is the line `save example, replace`. After converting to flongsep, for some reason we felt obligated to save the dataset. We will explain below. Now look farther down the history. After using _1_example.dta, we did not feel obligated to resave that dataset before using _2_example.dta. We will explain that below, too.

The flongsep style data are a matched set of datasets. You work with the $m = 0$ dataset in memory. It is your responsibility to save that dataset. Sometimes `mi` will have already saved the dataset for you. That was true here after `mi convert`, but it is impossible for you to know that in general, and it is your responsibility to save the dataset just as you would save any other dataset.

The $m > 0$ datasets, _#_name.dta, are `mi`'s responsibility. We do not have to concern ourselves with saving them. Obviously, it was not necessary to save them here because we had just used the data and made no changes. The point is that, in general, the $m > 0$ datasets are not our responsibility. The $m = 0$ dataset, however, is our responsibility.

We are done with the demonstration:

```
. drop _all
. mi erase example
(files example.dta _1_example.dta _2_example.dta erased)
```

How we constructed this example

You might be curious as to how we constructed `miproto.dta`. Here is what we did:

```
. drop _all
. input a b
             a         b
  1. 1 2
  2. 4 .
  3. end
. mi set wide
. mi set M = 2
(2 imputations added; M = 2)
```

```
. mi register regular a
. mi register imputed b
. replace _1_b = 4.5 in 2
(1 real change made)
. replace _2_b = 5.5 in 2
(1 real change made)
. mi passive: gen c = a + b
m=0:
(1 missing value generated)
m=1:
m=2:
. order a b c _1_b _2_b _1_c _2_c _mi_miss
```

Using mi system variables

You can use mi's system variables to make some tasks easier. For instance, if you wanted to know the overall number of complete and incomplete observations, you could type

```
. tabulate _mi_miss
```

because in all styles, the _mi_miss variable is created in $m = 0$ containing 0 if complete and 1 if incomplete.

If you wanted to know the summary statistics for weight in $m = 1$, the general solution is

```
. mi xeq 1: summarize weight
```

If you were using wide data, however, you could instead type

```
. summarize _1_weight
```

If you were using flong data, you could type

```
. summarize weight if _mi_m==1
```

If you were using mlong data, you could type

```
. summarize weight if (_mi_m==0 & !_mi_miss) | _mi_m==1
```

Well, that last is not so convenient.

What is convenient to do directly depends on the style you are using. Remember, however, you can always switch between styles by using mi convert (see [MI] **mi convert**). If you were using mlong data and wanted to compare summary statistics of the weight variable in the original data and in all imputations, you could type

```
. mi convert wide
. summarize *weight
```

Advice for using flongsep

Use the flongsep style when your data are too big to fit into any of the other styles. If you already have flongsep data, you can try to convert it to another style. If you get the error "no room to add more observations" or "no room to add more variables", then you need to increase the amount of memory dedicated to Stata (see help memory or see [D] **memory**) or resign yourself to using the flongsep style.

There is nothing wrong with the flongsep style except that you need to learn some new habits. Usually, in Stata, you work with a copy of the data in memory, and the changes you make are not reflected in the underlying disk file until and unless you explicitly save the data. If you want to change the name of the data, you merely save them in a file of a different name. None of that is true when working with flongsep data. Flongsep data are a collection of datasets; you work with the one corresponding to $m = 0$ in memory, and `mi` handles keeping the others in sync. As you make changes, the datasets on disk change.

Think of the collection of datasets as having one name. That name is established when the flongsep data are created. There are three ways that can happen. You might start with a non-`mi` dataset in memory and `mi set` it; you might import a dataset into Stata and the result be flongsep; or you might convert another `mi` dataset to flongsep. Here are all the corresponding commands:

. mi set flongsep *name*   (1)

. mi import flongsep *name*   (2)

. mi import nhanes1 *name*

. mi convert flongsep *name*   (3)

In each command, you specify a name and that name becomes the name of the flongsep dataset collection. In particular, *name*.dta becomes $m = 0$, _1_*name*.dta becomes $m = 1$, _2_*name*.dta becomes $m = 2$, and so on. You `use` flongsep data by typing `use` *name*, just as you would any other dataset. As we said, you work with $m = 0$ in memory and `mi` handles the rest.

Flongsep data are stored in the current (working) directory. Learn about `pwd` to find out where you are and about `cd` to change that; see [D] **cd**.

As you work with flongsep data, it is your responsibility to `save` *name*.dta almost as it would be with any Stata dataset. The difference is that `mi` might and probably has saved *name*.dta along the way without mentioning the fact, and `mi` has doubtlessly updated the _#_*name*.dta datasets, too. Nevertheless, it is still your responsibility to `save` *name*.dta when you are done because you do not know whether `mi` has saved *name*.dta recently enough. It is not your responsibility to worry about _#_*name*.dta.

It is a wonderful feature of Stata that you can usually work with a dataset in memory without modifying the original copy on disk except when you intend to update it. It is a unpleasant feature of flongsep that the same is not true. We therefore recommend working with a copy of the data, and `mi` provides an `mi copy` command for just that purpose:

. mi copy *newname*

With flongsep data in memory, when you type `mi copy` *newname*, the current flongsep files are saved in their existing name (this is one case where you are not responsible for saving *name*.dta), and then the files are copied to *newname*, meaning that $m = 0$ is copied to *newname*.dta, $m = 1$ is copied to _1_*newname*.dta, and so on. You are now working with the same data, but with the new name *newname*.

As you work, you may reach a point where you would like to save the data collection under *name* and continue working with *newname*. Do the following:

. mi copy *name*, replace

. use *newname*

When you are done for the day, if you want your data saved, do not forget to save them by using `mi copy`. It is also a good idea to erase the flongsep *newname* dataset collection:

. mi copy *name*, replace

. mi erase *newname*

By the way, *name*.dta, _1_*name*.dta, ... are just ordinary Stata datasets. By using general (non-mi) Stata commands, you can look at them and even make changes to them. Be careful about doing the latter; see [MI] **technical**.

See [MI] **mi copy** to learn more about `mi copy`.

Also see

[MI] **intro** — Introduction to mi

[MI] **mi copy** — Copy mi flongsep data

[MI] **mi erase** — Erase mi datasets

[MI] **technical** — Details for programmers

Title

> **technical** — Details for programmers

Description

Technical information for programmers who wish to extend mi is provided below.

Remarks

Remarks are presented under the following headings:

    Notation
    Definition of styles
        Style all
        Style wide
        Style mlong
        Style flong
        Style flongsep
        Style flongsep_sub
    Adding new commands to mi
    Outline for new commands
    Utility routines
        u_mi_assert_set
        u_mi_certify_data
        u_mi_no_sys_vars and u_mi_no_wide_vars
        u_mi_zap_chars

        u_mi_xeq_on_tmp_flongsep
        u_mi_get_flongsep_tmpname
        mata: u_mi_flongsep_erase()

        u_mi_sortback
        u_mi_save and u_mi_use
        mata: u_mi_wide_swapvars()

        u_mi_fixchars
        mata: u_mi_cpchars_get() and mata: u_mi_cpchars_put()
        mata: u_mi_get_mata_instanced_var()

        mata: u_mi_ptrace_*()
    How to write other set commands to work with mi

Notation

$$M = \text{\# of imputations}$$

$m = $ imputation number
    0. original data with missing values
    1. first imputation dataset
      .
      .
    M. last imputation dataset

$N = $ number of observations in $m = 0$

Definition of styles

Style describes how the mi data are stored. There are four styles: wide, mlong, flong, and flongsep.

Style all
Characteristics:
    _dta[_mi_marker]   "_mi_ds_1"

Description: _dta[_mi_marker] is set with all styles, including flongsep_sub. The definitions below apply only if "'_dta[_mi_marker]'" = "_mi_ds_1".

Style wide
Characteristics:
    _dta[_mi_style]   "wide"
    _dta[_mi_M] M
    _dta[_mi_ivars]   imputed variables; variable list
    _dta[_mi_pvars]   passive variables; variable list
    _dta[_mi_rvars]   regular variables; variable list
    _dta[_mi_update]   time last updated; %tc_value/1000

Variables:
    _mi_miss   whether incomplete; 0 or 1
    _#_varname   *varname* for $m = \#$, defined for each
                        '_dta[_mi_ivars]' and '_dta[_mi_pvars]'

Description: $m = 0, m = 1, \ldots, m = M$ are stored in one dataset with $\_N = N$ observations. Each imputed and passive variable has M additional variables associated with it. If variable bp contains the values in $m = 0$, then values for $m = 1$ are contained in variable _1_bp, values for $m = 2$ in _2_bp, and so on. wide stands for *wide*.

Style mlong
Characteristics:
    _dta[_mi_style]   "mlong"
    _dta[_mi_M] M
    _dta[_mi_N] N
    _dta[_mi_n]   # of observations in marginal
    _dta[_mi_ivars]   imputed variables; variable list
    _dta[_mi_pvars]   passive variables; variable list
    _dta[_mi_rvars]   regular variables; variable list
    _dta[_mi_update]   time last updated; %tc_value/1000

Variables:
    _mi_m m; $0, 1, \ldots, M$
    _mi_id   ID; $0, 1, \ldots, N$
    _mi_miss   whether incomplete; 0 or 1 if mi_m = 0, else .

Description: mlong: $m = 0, m = 1, \ldots, m = M$ are stored in one dataset with $\_N = N + M \times n$ observations, where n is the number of incomplete observations in $m = 0$. mlong stands for *marginal long*.

Style flong

Characteristics:

_dta[_mi_style]	"flong"
_dta[_mi_M]	M
_dta[_mi_N]	N
_dta[_mi_ivars]	imputed variables; variable list
_dta[_mi_pvars]	passive variables; variable list
_dta[_mi_rvars]	regular variables; variable list
_dta[_mi_update]	time last updated; %tc_value/1000

Variables:

_mi_m	m; 0, 1, ..., M
_mi_id	ID; 0, 1, ..., N
_mi_miss	whether incomplete; 0 or 1 if mi_m = 0, else .

Description: $m = 0$, $m = 1$, ..., $m = M$ are stored in one dataset with $\_N = N + M \times N$ observations, where N is the number of observations in $m = 0$. flong stands for *full long*.

Style flongsep

Characteristics:

_dta[_mi_style]	"flongsep"
_dta[_mi_name]	*name*
_dta[_mi_M]	M
_dta[_mi_N]	N
_dta[_mi_ivars]	imputed variables; variable list
_dta[_mi_pvars]	passive variables; variable list
_dta[_mi_rvars]	regular variables; variable list
_dta[_mi_update]	time last updated; %tc_value/1000

Variables:

_mi_id	ID; 0, 1, ..., N
_mi_miss	whether incomplete; 0 or 1

Description: $m = 0$, $m = 1$, ..., $m = M$ are each separate .dta datasets. If $m = 0$ data are stored in pat.dta, then $m = 1$ data are stored in _1_pat.dta, $m = 2$ in _2_pat.dta, and so on.

The definitions above apply only to $m = 0$, the dataset named '_dta[_mi_name]'.dta. See flongsep_sub directly below for $m > 0$. flongsep stands for *full long and separate*.

Style flongsep_sub

Characteristics:

_dta[_mi_style]	"flongsep_sub"
_dta[_mi_name]	*name*
_dta[_mi_m]	m; 0, 1, ..., M

Variables:

_mi_id	ID; 0, 1, ..., N

Description: The description above applies to the _'_dta[_mi_m]'_'_dta[_mi_name]'.dta datasets. There are M such datasets recording $m = 1$, ..., M used by the flongsep style directly above.

Adding new commands to mi

New commands are written in ado. Name the new command mi_cmd_*newcmd* and store it in mi_cmd_*newcmd*.ado. When the user types mi *newcmd* ..., mi_cmd_*newcmd*.ado will be executed.

Outline for new commands

```
program mi_cmd_newcmd, rclass                           (1)
        version 11
        u_mi_assert_set                                 (2)
        syntax ... [, ... noUPdate ...]                 (3)
        ...
        u_mi_certify_data, acceptable                   (4)
        ...
        if ("`update'"=="") {
                u_mi_certify_data, proper               (5)
        }
        ...
end
```

Notes:

1. The command may be rclass; that is not required. It may be eclass instead if you wish.

2. u_mi_assert_set verifies that the data are mi data; see *u_mi_assert_set* below.

3. If your command intends to use mi update to update the data before performing its intended task, include a noupdate option; see [MI] **noupdate option**. Some commands instead or in addition run mi update to perform cleanup after performing their task. Such use does not require a noupdate option.

4. u_mi_certify_data is the internal routine that performs mi update. An update is divided into two parts, called acceptable and proper. All commands should verify that the data are acceptable; see *u_mi_certify_data* below.

5. u_mi_certify_data, proper performs the second step of mi update; it verifies acceptable data are proper. Whether you verify properness is up to you but, if you do, you are supposed to include a noupdate option to skip running the check.

Utility routines

The only information you absolutely need to know is that already revealed. Using the utility routines described below, however, will simplify your programming task and make your code appear more professional to the end user.

As you read what follows, remember that you can review the source code for the routines by using viewsource; see [P] **viewsource**. If you wanted to see the source for u_mi_assert_set, you type viewsource u_mi_assert_set.ado. If you do this, you will sometimes see that the routines allow options not documented below. Ignore those options. They may not appear in future releases.

Using viewsource, you can also see examples of the utility commands being used by viewing the source of the mi commands we have written. Each mi command appears in the file mi_cmd_*command*.ado. Also remember that other mi commands make useful utility routines. For instance, if your new command makes passive variables, use mi register to register them. Always call existing mi commands through mi; code mi passive and not mi_cmd_passive.

u_mi_assert_set

    u_mi_assert_set [*desired_style*]

This utility verifies that data are mi and optionally of the desired style; it issues the appropriate error message and stops execution if not. The optional argument *desired_style* can be wide, mlong, flong, or flongsep, but is seldom specified. When not specified, any style is allowed.

u_mi_certify_data

    u_mi_certify_data [, acceptable proper noupdate sortok]

This command performs mi update. mi update is equivalent to u_mi_certify_data, acceptable proper sortok.

Specify one or both of acceptable and proper. If the noupdate option is specified, then proper is specified. The sortok option specifies that u_mi_certify_data need not spend extra time to preserve and restore the original sort order of the data.

An update is divided into two parts. In the first part, called acceptable, $m = 0$ and the _dta[_mi_*] characteristics are certified. Your program will use the information recorded in those characteristics and, before that information can be trusted, the data must be certified as acceptable. Do not trust any _dta[_mi_*] characteristics until you have run u_mi_certify_data, acceptable.

u_mi_certify_data, proper verifies that data known to be acceptable are proper. In practice, this means that in addition to trusting $m = 0$, you can trust $m > 0$.

Running u_mi_certify_data, acceptable might actually result in the data being certified as proper, although you cannot depend on that. When you run u_mi_certify_data, acceptable and certain problems are observed in $m = 0$, they are fixed in all m, and that can lead to other problems being detected, and by the time the whole process is through, the data are proper.

u_mi_no_sys_vars and u_mi_no_wide_vars

    u_mi_no_sys_vars "*variable_list*" ["*word*"]

    u_mi_no_wide_vars "*variable_list*" ["*word*"]

These routines are for use in parsing user input.

u_mi_no_sys_vars verifies that the specified list of variable names does not include any mi system variables such as _mi_m, _mi_id, _mi_miss, etc.

u_mi_no_wide_vars verifies that the specified list of variable names does not include any style wide $m > 0$ variables of the form _#_varname. u_mi_no_wide_vars may be called with any style of data but does nothing if the style is not wide.

Both functions issue appropriate error messages if problems are found. If *word* is specified, the error message will be "*word* may not include ...". Otherwise, the error message is "may not specify ...".

u_mi_zap_chars

    u_mi_zap_chars

u_mi_zap_chars deletes all _dta[_mi_*] characteristics from the data in memory.

u_mi_xeq_on_tmp_flongsep

    u_mi_xeq_on_tmp_flongsep [, nopreserve] : *command*

    u_mi_xeq_on_tmp_flongsep executes *command* on the data in memory, said data converted to style flongsep, and then converts the flongsep result back to the original style. If the data already are flongsep, a temporary copy is made and, at the end, posted back to the original. Either way, *command* is run on a temporary copy of the data. If anything goes wrong, the user's original data are restored. That is, they are restored unless nopreserve is specified. If *command* completes without error, the flongsep data in memory are converted back to the original style and the original data discarded.

    It is not uncommon to write commands that can deal only with flongsep data, and yet these seem to users as if they work with all styles. That is because the routines use u_mi_xeq_on_tmp_flongsep. They start by allowing any style, but the guts of the routine are written assuming flongsep. mi stjoin is implemented in this way. There are two parts to mi stjoin: mi_cmd_stjoin.ado and mi_sub_stjoin_flongsep.ado. mi_cmd_stjoin.ado ends with

    u_mi_xeq_on_tmp_flongsep: mi_sub_stjoin_flongsep 'if', 'options'

    mi_sub_stjoin_flongsep does all the work, while u_mi_xeq_on_tmp_flongsep handles the issue of converting to flongsep and back again. The mi_sub_stjoin_flongsep subroutine must appear in its own ado-file because u_mi_xeq_on_tmp_flongsep is itself implemented as an ado-file. u_mi_xeq_on_tmp_flongsep would be unable to find the subroutine otherwise.

u_mi_get_flongsep_tmpname

    u_mi_get_flongsep_tmpname *macname* : *basename*

    u_mi_get_flongsep_tmpname creates a temporary flongsep name based on *basename* and stores it in the local macro *macname*. u_mi_xeq_on_tmp_flongsep, for your information, obtains the temporary name it uses from this routine.

    u_mi_get_flongsep_tmpname is seldom used directly because u_mi_xeq_on_tmp_flongsep works well for shifting temporarily into flongsep mode, and u_mi_xeq_on_tmp_flongsep does a lot more than just getting a name under which the data should be temporarily stored. There are instances, however, when one needs to be more involved in the conversion. For examples, see the source mi_cmd_append.ado and mi_cmd_merge.ado. The issue these two routines face is that they need to shift two input datasets to flongsep, then they create a third from them, and that is the only one that needs to be shifted back to the original style. So these two commands handle the conversions themselves using u_mi_get_flongsep_tmpname and mi convert.

    For instance, they start with something like

    u_mi_get_flongsep_tmpname master : __mimaster

That creates a temporary name suitable for use with mi convert and stores it in 'master'. The suggested name is __mimaster, but if that name is in use, u_mi_get_flongsep_tmpname will form from it __mimaster1, or __mimaster2, etc. It is recommended that you specify a *basename* that begins with __mi, which is to say, two underscores followed by mi.

    Next you must appreciate that it is your responsibility to eliminate the temporary files. You do that by coding something like

```
        ...
        local origstyle "`_dta[_mi_style]'"
        if ("`origstyle'"=="flongsep") {
                local origstyle "`origstyle' `_dta[_mi_name]'"
        }
        u_mi_get_flongsep_tmpname master : __mimaster
        capture {
                quietly mi convert flongsep `master'
                ...
                ...
                quietly mi convert `origstyle', clear replace
        {
        nobreak {
                local rc = _rc
                mata: u_mi_flongsep_erase("`master'", 0, 0)
                if (`rc') {
                        exit `rc'
                }
        }
}
```

The other thing to note above is our use of `mi convert 'master'` to convert our data to flongsep under the name `'master'`. What, you might wonder, happens if our data already is flongsep? A nice feature of `mi convert` is that, run on already flongsep data, it performs an `mi copy`; see [MI] **mi copy**.

mata: u_mi_flongsep_erase()

mata: u_mi_flongsep_erase("*name*", *from* [, *output*])

where

	name	*string*; flongsep name
	from	#; where to begin erasing
	output	0|1; whether to produce output

`mata: u_mi_flongsep_erase()` is the internal version of `mi erase` (see [MI] **mi erase**); use whichever is more convenient.

Input *from* is usually specified as 0 and then `mata: u_mi_flongsep_erase()` erases *name*.dta, _1_*name*.dta, _2_*name*.dta, and so on. *from* may be specified as a number greater than zero, however, and then erased are _<*from*>_*name*.dta, _<*from*+1>_*name*.dta, _<*from*+2>_*name*.dta, ...

If *output* is 0, no output is produced; otherwise, the erased files are also listed. If *output* is not specified, files are listed.

See `viewsource u_mi.mata` for the source code for this routine.

u_mi_sortback

u_mi_sortback *varlist*

`u_mi_sortback` removes dropped variables from *varlist* and sorts the data on the remaining variables. The routine is for dealing with sort-preserve problems when `program` *name*, `sortpreserve` is not adequate, such as when the data might be subjected to substantial editing between the preserving of the sort order and the restoring of it. To use `u_mi_sortback`, first record the order of the data:

technical — Details for programmers

```
local sortedby : sortedby
tempvar recnum
gen long 'recnum' = _n
quietly compress 'recnum'
```

Later, when you want to restore the sort order, you code

```
u_mi_sortback 'sortedby' 'recnum'
```

u_mi_save and u_mi_use

u_mi_save *macname* : *filename* [, *save_options*]

u_mi_use '"'*macname*'"' *filename* [, clear <u>nol</u>abel]

save_options are as described in [D] **save**. clear and nolabel are as described in [D] **use**. In both commands, *filename* must be specified in quotes if it contains any special characters or blanks.

It is sometimes necessary to save data in a temporary file and reload them later. In such cases, when the data are reloaded, you would like to have the original c(filename), c(filedate), and c(changed) restored. u_mi_save saves that information in *macname*. u_mi_use restores the information from the information saved in *macname*. Note the use of compound quotes around '*macname*' in u_mi_use; they are not optional.

mata: u_mi_wide_swapvars()

mata: u_mi_wide_swapvars(*m*, *tmpvarname*)

where

$$m \quad\quad \#;\ 1 \le \# \le M$$
$$tmpvarname \quad\quad string;\ \text{name from tempvar}$$

This utility is for use with wide data only. For each variable name contained in _dta[_mi_ivars] and _dta[_mi_pvars], mata: u_mi_wide_swapvars() swaps the contents of *varname* with _*m*_*varname*. Argument *tmpvarname* must be the name of a temporary variable obtained from command tempvar, and the variable must not exist. mata: u_mi_wide_swapvars() will use this variable while swapping. See [P] **macro** for more information on tempvar.

This function is its own inverse, assuming _dta[_mi_ivars] and _dta[_mi_pvars] have not changed.

See viewsource u_mi.mata for the source code for this routine.

u_mi_fixchars

u_mi_fixchars [, acceptable proper]

u_mi_fixchars makes the data and variable characteristics the same in $m = 1$, $m = 2$, ..., $m = M$ as they are in $m = 0$. The options specify what is already known to be true about the data, that the data are known to be acceptable or known to be proper. If neither is specified, you are stating that you do not know whether the data are even acceptable. That is okay. u_mi_fixchars handles performing whatever certification is required. Specifying the options makes u_mi_fixchars run faster.

This stabilizing of the characteristics is not about mi's characteristics; that is handled by u_mi_certify_data. Other commands of Stata set and use characteristics, and u_mi_fixchars ensures that those characteristics are the same across all m.

mata: u_mi_cpchars_get() and mata: u_mi_cpchars_put()

>   mata: u_mi_cpchars_get(*matavar*)

>   mata: u_mi_cpchars_put(*matavar*, {0 | 1 | 2})

where *matavar* is a Mata transmorphic variable. Obtain *matavar* from u_mi_get_mata_instanced_var() when using these functions from Stata.

These routines replace the characteristics in one dataset with those of another. They are used to implement u_mi_fixchars.

mata: u_mi_cpchars_get(*matavar*) stores in *matavar* the characteristics of the data in memory. The data in memory remain unchanged.

mata: u_mi_cpchars_put(*matavar*, #) replaces the characteristics of the data in memory with those previously recorded in *matavar*. The second argument specifies the treatment of _dta[_mi_*] characteristics:

>   0   delete them in the destination data
>   1   copy them from the source just like any other characteristic
>   2   retain them as-is from the destination data.

mata: u_mi_get_mata_instanced_var()

>   mata: u_mi_get_mata_instanced_var("*macname*", "*basename*" [, *i_value*])

where

>   *macname*   name of local macro
>   *basename*   suggested name for instanced variable
>   *i_value*   initial value for instanced variable

mata: u_mi_get_mata_instanced_var() creates a new Mata global variable, initializes it with *i_value* or as a 0×0 real, and places its name in local macro *macname*. Typical usage is

```
local var
capture noisily {
        mata: u_mi_get_mata_instanced_var("var", "myvar")
        ...
        ... use `var' however you wish ...
        ...
}
nobreak {
        local rc = _rc
        capture mata: mata drop `var'
        if (`rc') {
                exit `rc'
        }
}
```

mata: u_mi_ptrace_*()

h = u_mi_ptrace_open("*filename*", {"r" | "w"} [, {0|1}])

u_mi_ptrace_write_stripes(h, *id*, *ynames*, *xnames*)

u_mi_ptrace_write_iter(h, *m*, *iter*, *B*, *V*)

u_mi_ptrace_close(h)

u_mi_ptrace_safeclose(h)

The above are Mata functions, where

>*h*, if it is declared, should be declared transmorphic
>*id* is a string scalar
>*ynames* and *xnames* are string scalars
>*m* and *iter* are real scalars
>*B* and *V* are real matrices; *V* must be symmetric

These routines write parameter-trace files; see [MI] **mi ptrace**. The procedure is 1) open the file; 2) write the stripes; 3) repeatedly write iteration information; and 4) close the file.

1. Open the file: *filename* may be specified with or without a file suffix. Specify the second argument as "w". The third argument should be 1 if the file may be replaced when it exists, and 0 otherwise.

2. Write the stripes: Specify *id* as the name of your routine or as ""; mi ptrace describe will show this string as the creator of the file if the string is not "". *ynames* and *xnames* are both string scalars containing space-separated names or, possibly, *op.names*.

3. Repeatedly write iteration information: Written are m, the imputation number; *iter*, the iteration number; B, the matrix of coefficients; and V, the variance matrix. B must be $ny \times nx$ and V must be $ny \times ny$ and symmetric, where nx = length(tokens(*xnames*)) and ny = length(tokens(*ynames*)).

4. Close the file: In Mata, use u_mi_ptrace_close(h). It is highly recommended that, before step 1, h be obtained from inside Stata (not Mata) using mata: u_mi_get_mata_instanced_var("h", "*myvar*"). If you follow this advice, include a mata: u_mi_ptrace_safeclose('h') in the ado-file cleanup code. This will ensure that open files are closed if the user presses *Break* or something else causes your routine to exit before the file is closed. A correctly written program will have two closes, one in Mata and another in the ado-file, although you could omit the one in Mata. See *mata: u_mi_get_mata_instanced_var()* directly above.

Also included in u_mi_ptrace_*() are routines to read parameter-trace files. You should not need these routines because users will use Stata command mi ptrace use to load the file you have written. If you are interested, however, then viewsource u_mi_ptrace.mata.

How to write other set commands to work with mi

This section concerns the writing of other set commands such as [ST] **stset** or [XT] **xtset**—set commands having nothing to do with mi—so that they properly work with mi.

The definition of a set command is any command that creates characteristics in the data, and possibly creates variables in the data, that other commands in the suite will subsequently access. Making such set commands work with mi is mostly mi's responsibility, but there is a little you need to do to assist mi. Before dealing with that, however, write and debug your set command ignoring mi. Once that is done, go back and add a few lines to your code. We will pretend your set command is named mynewset and your original code looks something like this:

```
program mynewset
    ...
    syntax ... [, ... ]
    ...
end
```

Our goal is to make it so that mynewset will not run on mi data while simultaneously making it so that mi can call it (the user types mi mynewset). When the user types mi mynewset, mi will 1) give mynewset a clean, $m = 0$ dataset on which it can run and 2) duplicate whatever mynewset does to $m = 0$ on $m = 1, m = 2, \ldots, m = M$.

To achieve this, modify your code to look like this:

```
program mynewset
    ...
    syntax ... [, ... MI]                                      (1)
    if ("`mi'"=="") {                                          (2)
            u_mi_not_mi_set "mynewset"
            local checkvars "*"                                (3)
    }
    else {
            local checkvars "u_mi_check_setvars settime"       (3)
    }
    ...
    `checkvars' `varlist'                                      (4)
    ...
end
```

That is,

1. Add the mi option to any options you already have.

2. If the mi option is not specified, execute u_mi_not_mi_set, passing to it the name of your set command. If the data are not mi, then u_mi_not_mi_set will do nothing. If the data are mi, then u_mi_not_mi_set will issue an error telling the user to run mi mynewset.

3. Set new local macro checkvars to * if the mi option is not specified, and otherwise to u_mi_check_setvars. We should mention that the mi option will be specified when mi mynewset calls mynewset.

4. Run 'checkvars' on any input variables mynewset uses that must not vary across m. mi does not care about other variables or even about new variables mynewset might create; it cares only about existing variables that should not vary across m.

    Let's understand what "'checkvars' *varlist*" does. If the mi option was not specified, the line expands to "* *varlist*", which is a comment, and does nothing. If the mi option was specified, the line expands to "u_mi_check_setvars settime *varlist*". We are calling mi routine u_mi_check_setvars, telling it that we are calling at set time, and passing along *varlist*. u_mi_check_setvars will verify that *varlist* does not contain mi system variables or variables that vary across m. Within mynewset, you may call 'checkvars' repeatedly if that is convenient.

You have completed the changes to mynewset. You finally need to write one short program that reads

```
program mi_cmd_mynewset
        version 11
        mi_cmd_genericset '"mynewset '0'"' "_mynewset_x _mynewset_y"
end
```

In the above, we assume that mynewset might add one or two variables to the data named _mynewset_x and _mynewset_y. List in the second argument all variables mynewset might create. If mynewset never creates new variables, then the program should read

```
program mi_cmd_mynewset
        version 11
        mi_cmd_genericset '"mynewset '0'"'
end
```

You are done.

Also see

[MI] **intro** — Introduction to mi

Title

workflow — Suggested workflow

Description

Provided below are suggested workflows for working with original data and for working with data already having imputations.

Remarks

Remarks are presented under the following headings:

> Suggested workflow for original data
> Suggested workflow for data already having imputations
> Example

Suggested workflow for original data

By original data, we mean data with missing values for which you do not already have imputations. Your task is to identify the missing values, impute values for them, and perform estimation.

mi does not have a fixed order in which you must perform tasks except that you must mi set the data first.

1. mi set your data; see [MI] **mi set**.

    Set the data to be wide, mlong, flong, or flongsep. Choose flongsep only if your data are bumping up against the constraints of memory. Choose flong or flongsep if you will need super-varying variables.

    Memory is not usually a problem, and super-varying variables are seldom necessary, so we generally start with the data as wide:

    . use *originaldata*

    . mi set wide

    If you need to use flongsep, you also need to specify a name for the flongsep dataset collection. Choose a name different from the current name of the dataset:

    . use *originaldata*

    . mi set flongsep *newname*

    If the original data is chd.dta, you might choose chdm for *newname*. *newname* does not include the .dta suffix. If you choose chdm, the data will then be stored in chdm.dta, _1_chdm.dta, and so on. That is why it is important that you choose a name different from *originaldata*; you do not want your mi data to overwrite the original. Stata users are used to working with a copy of the data in memory, meaning that the changes made to the data are not reflected in the .dta dataset until the user saves them. With flongsep data, however, changes are made to the mi .dta dataset collection as you work. See *Advice for using flongsep* in [MI] **styles**.

2. Use `mi describe` often; see [MI] **mi describe**.

   `mi describe` will not tell you anything useful yet, but as you set more about the data, `mi describe` will be more informative.

   ```
   . mi describe
   ```

3. Use `mi misstable` to identify missing values; see [MI] **mi misstable**.

   `mi misstable` is just the standard `misstable` (see [R] **misstable**), but tailored for `mi` data. Several Stata commands have `mi` variants. Become familiar with them. If there is no `mi` variant, then it is generally safe to use the standard command directly, although it may not be appropriate. For instance, typing `misstable` rather than `mi misstable` would produce appropriate results right now, but it would not produce appropriate results later. If `mi` datasets $m = 0$, $m = 1$, ..., $m = M$ exist and you run `misstable`, you might end up running the command on a strange combination of the m's. We recommend the wide style because general Stata commands will do what you expect. The same is true for the flongsep style. It is your responsibility to get this right.

   So what is the difference between `mi misstable` and `misstable`? `mi misstable` amounts to `mi xeq 0: misstable, exok`, which is to say, it runs on $m = 0$ and specifies the `exok` option so that extended missing values are treated as hard missings.

   In general, you need to become familiar with all the `mi` commands, use the `mi` variant of regular Stata commands whenever one exists, and think twice before using a command without an `mi` prefix. Doing the right thing will become automatic once you gain familiarity with the styles; see [MI] **styles**.

   To learn about the missing values in your data, type

   ```
   . mi misstable summarize
   ```

4. Use `mi register imputed` to register the variables you wish to impute; see [MI] **mi set**.

   The only variables that `mi` will impute are those registered as `imputed`. You can register variables one at a time or all at once. If you register a variable mistakenly, use `mi unregister` to unregister it.

   ```
   . mi register imputed varname [varname ...]
   ```

5. Use `mi impute` to impute (fill in) the missing values; see [MI] **mi impute**.

   There is a lot to be said here. For instance, in a dataset where variables `age` and `smokes` contain missing, you might type

   ```
   . mi register age bmi
   . mi impute mvn age bmi = attack smokes hsgrad, add(10)
   ```

   `mi impute`'s `add(#)` option specifies the number of imputations to be added. We currently have 0 imputations, so after imputation, we will have 10. We usually start with a small number of imputations and add more later.

6. Use `mi describe` to verify that all missing values are filled in; see [MI] **mi describe**.

   ```
   . mi describe
   ```

   You might also want to use `mi xeq` (see [MI] **mi xeq**) to look at summary statistics in each of the imputation datasets:

   ```
   . mi xeq: summarize
   ```

7. Generate passive variables; see [MI] **mi passive**.

   Passive variables are variables that are functions of imputed variables, such as lnage when some values of age are imputed. The values of passive variables differ across m just as the values of imputed variables do. The official way to generate imputed values is by using mi passive:

   . mi passive: generate lnage = ln(age)

   Rather than use the official way, however, we often switch our data to mlong and just generate the passive variables directly:

   . mi convert mlong
   . generate lnage = ln(age)
   . mi register passive lnage

   If you work as we do, remember to register any passive variables you create. When you are done, you can mi convert your data back to wide, but there is no reason to do that.

8. Use mi estimate (see [MI] **mi estimate**) to fit models:

   . mi estimate: logistic attack smokes age bmi hsgrad

   You fit your model just as you would ordinarily, except that you add mi estimate: in front of the command.

To see an example of the advice applied to a simple dataset, see *Example* below.

In theory, you should get your data cleaning and data management out of the way before mi setting your data. In practice, that will not happen, and so you will want to become familiar with the other mi commands. Among the data-management commands available are mi append (see [MI] **mi append**), mi merge (see [MI] **mi merge**), mi expand (see [MI] **mi expand**), and mi reshape (see [MI] **mi reshape**. If you are working with survival-time data, also see [MI] **mi stsplit**. To stset your data, or svyset, or xtset, see [MI] **mi set** and [MI] **mi XXXset**.

Suggested workflow for data already having imputations

Data sometimes come with imputations included. The data might be made by another researcher for you or the data might come from an official source. Either way, we will assume that the data are not in Stata format, because if they were, you would just use the data and would type mi describe.

mi can import officially produced datasets created by the National Health and Nutrition Examination Survey (NHANES) with the mi import nhanes1 command, and mi can import more informally created datasets that are wide-, flong-, or flongsep-like with mi import wide, mi import flong, or mi import flongsep; see [MI] **mi import**.

The required workflow is hardly different from *Suggested workflow for original data*, presented above. The differences are that you will use mi import rather than mi set, and you will skip using mi impute to generate the imputations. In this sense, your job is easier.

On the other hand, you need to verify that you have imported your data correctly, and we have a lot to say about that. Basically, after importing, you need to be careful about which mi commands you use until you have verified that you have the variables registered correctly. That is discussed in [MI] **mi import**.

Example

We are going to repeat *A simple example* from [MI] **intro**, but this time we are going to follow the advice given above in *Suggested workflow for original data*.

We have fictional data on 154 patients and want to examine the relationship between binary outcome `attack`, recording heart attacks, and variables `smokes`, `age`, `bmi`, `hsgrad`, and `female`. We will use logistic regression. Below we load our original data and show you a little about it using the standard commands `describe` and `summarize`. We emphasize that `heart5.dta` is just a standard Stata dataset. It has not been `mi set`.

```
. use http://www.stata-press.com/data/r11/mheart5
(Fictional heart attack data, bmi and age missing)
. describe
Contains data from http://www.stata-press.com/data/r11/mheart5.dta
  obs:           154                          Fictional heart attack data; bmi
                                                and age missing
 vars:             6                          19 Jun 2009 10:50
 size:         2,464 (99.9% of memory free)
```

variable name	storage type	display format	value label	variable label
attack	byte	%9.0g		Outcome (heart attack)
smokes	byte	%9.0g		Current smoker
age	float	%9.0g		Age, in years
bmi	float	%9.0g		Body Mass Index, kg/m^2
female	byte	%9.0g		Gender
hsgrad	byte	%9.0g		High school graduate

Sorted by:

```
. summarize
```

Variable	Obs	Mean	Std. Dev.	Min	Max
attack	154	.4480519	.4989166	0	1
smokes	154	.4155844	.4944304	0	1
age	142	56.43324	11.59131	20.73613	83.78423
bmi	126	25.23523	4.029325	17.22643	38.24214
female	154	.2467532	.4325285	0	1
hsgrad	154	.7532468	.4325285	0	1

The first guideline is

1. `mi set` your data; see [MI] **mi set**.

We will set the data to be flong even though in *A simple example* we set the data to be mlong. `mi` provides four styles—flong, mlong, wide, and flongsep—and at this point it does not matter which we choose. `mi` commands work the same way regardless of style. Four styles are provided because, should we decide to step outside of `mi` and attack the data with standard Stata commands, we will find different styles more convenient depending on what we want to do. It is easy to switch styles.

Below we type `mi set flong` and then, to show you what that command did to the data, we show you the output from a standard `describe`:

```
. mi set flong

. describe
Contains data from http://www.stata-press.com/data/r11/mheart5.dta
   obs:           154                          Fictional heart attack data; bmi
                                                 and age missing
  vars:             9                          19 Jun 2009 10:50
  size:         3,234 (99.9% of memory free)
```

variable name	storage type	display format	value label	variable label
attack	byte	%9.0g		Outcome (heart attack)
smokes	byte	%9.0g		Current smoker
age	float	%9.0g		Age, in years
bmi	float	%9.0g		Body Mass Index, kg/m^2
female	byte	%9.0g		Gender
hsgrad	byte	%9.0g		High school graduate
_mi_miss	byte	%8.0g		
_mi_m	int	%8.0g		
_mi_id	int	%12.0g		

```
Sorted by:
```

Typing `mi set flong` added three variables to our data: _mi_miss, _mi_m, and _mi_id. Those variables belong to mi. If you are curious about them, see [MI] **styles**. Advanced users can even use them. No matter how advanced you are, however, you must never change their contents.

Except for the three added variables, the data are unchanged, and we would see that if we typed `summarize`. The three added variables are due to the style we chose. When you `mi set` your data, different styles will change the data differently, but the changes will be just around the edges.

The second guideline is

2. Use `mi describe` often; see [MI] **mi describe**.

The guideline is to use `mi describe`, not `describe` as we just did. Here is the result:

```
. mi describe
  Style: flong
         last mi update 21jun2009 12:26:02, 5 seconds ago
  Obs.:  complete            154
         incomplete            0  (M = 0 imputations)
         ─────────────────────────
         total               154
  Vars.: imputed: 0
         passive: 0
         regular: 0
         system:  3; _mi_m _mi_id _mi_miss
         (there are 6 unregistered variables)
```

As the guideline warned us, "mi describe will not tell you anything useful yet."

The third guideline is

3. Use `mi misstable` to identify missing values; see [MI] **mi misstable**.

Below we type `mi misstable summarize` and `mi misstable nested`:

```
. mi misstable summarize
                                                        Obs<.
                                              Unique
        Variable      Obs=.      Obs>.   Obs<.  values      Min        Max

             age         12                 142     142  20.73613   83.78423
             bmi         28                 126     126  17.22643   38.24214

. mi misstable nested
   1.  age(12) -> bmi(28)
```

`mi misstable summarize` reports the variables containing missing values. Those variables in our data are `age` and `bmi`. Notice that `mi misstable summarize` draws a distinction between, as it puts it, "Obs=." and "Obs>.", which is to say, between standard missing (.) and extended missing (.a, .b, ..., .z). That is because `mi` has a concept of soft and hard missing, and it associates soft missing with system missing and hard missing with extended missing. Hard missing values—extended missings—are taken to mean missing values that are not to be imputed. Our data have no missing values like that.

After typing `mi misstable summarize`, we typed `mi misstable nested` because we were curious whether the missing values were nested or, to use the jargon, monotone. We discovered that they were. That is, `age` has 12 missing values in the data, and in every observation in which `age` is missing, so is `bmi`, although `bmi` has another 16 missing values scattered around the data. That means we can use a monotone imputation method, and that is good news because monotone methods are more flexible and faster. We will discuss the implications of that shortly. There is a mechanical detail we must handle first.

The fourth guideline is

4. Use `mi register imputed` to register the variables you wish to impute; see [MI] **mi set**.

We know that `age` and `bmi` have missing values and, before we can impute replacements for those missing values, we must register the variables as to-be-imputed, which we do by typing

```
. mi register imputed age bmi
(28 m=0 obs. now marked as incomplete)
```

Guideline 2 suggested that we type `mi describe` often. Perhaps now would be a good time:

```
. mi describe
  Style:  flong
          last mi update 21jun2009 12:26:02, 40 seconds ago
  Obs.:   complete        126
          incomplete       28   (M = 0 imputations)

          total           154
  Vars.:  imputed:  2;  age(12) bmi(28)
          passive:  0
          regular:  0
          system:   3;  _mi_m _mi_id _mi_miss
          (there are 4 unregistered variables; attack smokes female hsgrad)
```

The output has indeed changed. mi knows just as it did before that we have 154 observations, and it now knows that 126 of them are complete and 28 of them are incomplete. It also knows that age and bmi are to be imputed. The numbers in parentheses are the number of missing values.

The fifth guideline is

5. Use mi impute to impute (fill in) the missing values; see [MI] **mi impute**.

In *A simple example* from [MI] **intro**, we imputed values for age and bmi by typing

. mi impute mvn age bmi = attack smokes hsgrad female, add(10)

This time, we will impute values by typing

. mi impute monotone (regress) age bmi = attack smokes hsgrad female, add(20)

We changed add(10) to add(20) for no other reason than to show that we could, although we admit to a preference for more imputations whenever possible. add() specifies the number of imputations to be added to the data. For every missing value, we will impute 20 nonmissing replacements.

We switched from mi impute mvn to mi impute monotone because our data are monotone. Here mi impute monotone will be faster than mi impute mvn but will offer no statistical advantage. In other cases, there might be statistical advantages. All of which is to say that when you get to the imputation step, you have important decisions to make and you need to become knowledgeable about the subject. You can start by reading [MI] **mi impute**.

```
. set seed 20039
. mi impute monotone (regress) age bmi = attack smokes hsgrad female, add(20)
Conditional models:
          age: regress age attack smokes hsgrad female
          bmi: regress bmi age attack smokes hsgrad female

Multivariate imputation                     Imputations =       20
Monotone method                                   added =       20
Imputed: m=1 through m=20                       updated =        0

          age: linear regression
          bmi: linear regression
```

	Observations per *m*			
Variable	complete	incomplete	imputed	total
age	142	12	12	154
bmi	126	28	28	154

(complete + incomplete = total; imputed is the minimum across *m* of the number of filled in observations.)

Note that we typed set seed 20039 before issuing the mi impute command. Doing that made our results reproducible. We could have specified mi impute's rseed(20039) option instead. Or we could have skipped setting the random-number seed altogether, and then we would not be able to reproduce our results.

The sixth guideline is

6. Use mi describe to verify that all missing values are filled in; see [MI] **mi describe**.

```
. mi describe, detail
    Style:  flong
            last mi update 21jun2009 12:40:01, 2 seconds ago
    Obs.:   complete           126
            incomplete          28   (M = 20 imputations)
                            ───────
            total              154
    Vars.:  imputed: 2; age(12; 20*0) bmi(28; 20*0)
            passive: 0
            regular: 0
            system:  3; _mi_m _mi_id _mi_miss
            (there are 4 unregistered variables; attack smokes female hsgrad)
```

This time, we specified mi describe's detail option, although you have to look closely at the output to see the effect. When you do not specify detail, mi describe reports results for the original, unimputed data only, what we call $m = 0$ throughout this documentation. When you specify detail, mi describe also includes information about the imputation data, what we call $m > 0$ and is $m = 1, m = 2, \ldots, m = 20$ here. Previously, mi describe reported "age(12)", meaning that age in $m = 0$ has 12 missing values. This time, it reports "age(12; 20*0)", meaning that age still has 12 missing values in $m = 0$, and it has 0 missing values in the 20 imputations. bmi also has 0 missing values in the imputations. Success!

Let's take a detour to see how our data really look. Let's type Stata's standard describe command. The last time we looked, our data had three extra variables.

```
. describe
Contains data from http://www.stata-press.com/data/r11/mheart5.dta
  obs:         3,234                          Fictional heart attack data; bmi
                                                and age missing
 vars:             9                          21 Jun 2009 13:36
 size:        67,914 (99.4% of memory free)
─────────────────────────────────────────────────────────────────────────────
              storage  display     value
variable name   type   format      label      variable label
─────────────────────────────────────────────────────────────────────────────
attack          byte   %9.0g                  Outcome (heart attack)
smokes          byte   %9.0g                  Current smoker
age             float  %9.0g                  Age, in years
bmi             float  %9.0g                  Body Mass Index, kg/m^2
female          byte   %9.0g                  Gender
hsgrad          byte   %9.0g                  High school graduate
_mi_id          int    %12.0g
_mi_miss        byte   %8.0g
_mi_m           int    %8.0g
─────────────────────────────────────────────────────────────────────────────
Sorted by:
```

Nothing has changed as far as variables are concerned, but notice the number of observations. Previously, we had 154 observations. Now we have 3,234! That works out to 21*154. Stored is our original data plus 20 imputations. The flong style makes extra copies of the data.

We chose style flong only because it is so easy to explain. In *A simple example* from [MI] **intro** using this same data, we choose style mlong. It is not too late:

```
. mi convert mlong
```

All that is required to change styles is typing mi convert. The style of the data changes, but not the contents. Let's see what describe has to report:

```
. describe

Contains data from http://www.stata-press.com/data/r11/mheart5.dta
  obs:           714                          Fictional heart attack data; bmi
                                                and age missing
 vars:             9                          21 Jun 2009 13:36
 size:        14,994 (99.9% of memory free)
```

variable name	storage type	display format	value label	variable label
attack	byte	%9.0g		Outcome (heart attack)
smokes	byte	%9.0g		Current smoker
age	float	%9.0g		Age, in years
bmi	float	%9.0g		Body Mass Index, kg/m^2
female	byte	%9.0g		Gender
hsgrad	byte	%9.0g		High school graduate
_mi_id	int	%12.0g		
_mi_miss	byte	%8.0g		
_mi_m	int	%8.0g		

```
Sorted by:  _mi_m  _mi_id
```

The data look much like they did when they were flong, except that the number of observations has fallen from 3,234 to 714! Style mlong is an efficient style in that, rather than storing the full data for every imputation, it stores only the changes. Back when the data were flong, mi describe reported that we had 28 incomplete observations. We get 714 from the 154 original observations plus 20×28 replacement observations for the incomplete observations.

We recommend style mlong. Style wide is also recommended. Below we type mi convert to convert our mlong data to wide, and then run the standard describe command:

```
. mi convert wide

. describe

Contains data from http://www.stata-press.com/data/r11/mheart5.dta
  obs:           154                          Fictional heart attack data; bmi
                                                and age missing
 vars:            47                          21 Jun 2009 13:43
 size:        27,258 (99.7% of memory free)
```

variable name	storage type	display format	value label	variable label
attack	byte	%9.0g		Outcome (heart attack)
smokes	byte	%9.0g		Current smoker
age	float	%9.0g		Age, in years
bmi	float	%9.0g		Body Mass Index, kg/m^2
female	byte	%9.0g		Gender
hsgrad	byte	%9.0g		High school graduate
_mi_miss	byte	%8.0g		
_1_age	float	%9.0g		Age, in years
_1_bmi	float	%9.0g		Body Mass Index, kg/m^2
_2_age	float	%9.0g		Age, in years
_2_bmi	float	%9.0g		Body Mass Index, kg/m^2
(output omitted)				
_20_age	float	%9.0g		Age, in years
_20_bmi	float	%9.0g		Body Mass Index, kg/m^2

```
Sorted by:
```

In the wide style, our data are back to having 154 observations, but now we have 47 variables!

Variable _1_age contains age for $m = 1$, _1_bmi contains bmi for $m = 1$, _2_age contains age for $m = 2$, and so on.

Guideline 7 is

7. Generate passive variables.

Passive variables are variables derived from imputed variables. For instance, if we needed lnage = ln(age), variable lnage would be passive. Passive variables are easy to create; see [MI] **mi passive**. We are not going to need any passive variables in this example.

Guideline 8 is

8. Use mi estimate to fit models; see [MI] **mi estimate**.

Our data are wide right now, but that does not matter. We fit our model:

```
. mi estimate: logistic attack smokes age bmi hsgrad female
Multiple-imputation estimates              Imputations     =         20
Logistic regression                        Number of obs   =        154
                                           Average RVI     =     0.0854
DF adjustment:   Large sample              DF:      min    =     286.81
                                                    avg    =  158565.80
                                                    max    =  811336.09
Model F test:       Equal FMI              F(   5,10540.4) =       3.17
Within VCE type:          OIM              Prob > F        =     0.0074
```

attack	Coef.	Std. Err.	t	P>|t|	[95% Conf. Interval]	
smokes	1.165097	.3590394	3.25	0.001	.4613588	1.868835
age	.0299019	.0162879	1.84	0.067	-.0020498	.0618536
bmi	.0983207	.0508291	1.93	0.054	-.0017247	.198366
hsgrad	.1439488	.4031188	0.36	0.721	-.646158	.9340557
female	-.0936945	.4121954	-0.23	0.820	-.9015838	.7141949
_cons	-4.956249	1.772269	-2.80	0.005	-8.441389	-1.471108

Those familiar with the logistic command will be surprised that mi estimate: logistic reported coefficients rather than odds ratios. That is because the estimation command is not logistic using mi estimate, it is mi estimate using logistic. If we wanted to see odds ratios at estimation time, we could have typed

        . mi estimate, or: logistic ...

By the same token, if we wanted to replay results, we would not type logistic, we would type mi estimate:

        . mi estimate
          (output omitted)

If we wanted to replay results with odds ratios, we would type

        . mi estimate, or

And that concludes the guidelines.

Also see

[MI] **intro** — Introduction to mi

[MI] **Glossary**

Title

Glossary

Description

Please read. The terms defined below are used throughout the documentation, and sometimes without explanation.

Glossary

arbitrary missing pattern. Any missing-value pattern. Some imputation methods are suitable only when the pattern of missing values is special, such as a monotone-missing pattern. An imputation method suitable for use with an arbitrary missing pattern may be used regardless of the pattern.

casewise deletion. See *listwise deletion*.

complete and incomplete observations. An observation in the $m=0$ data is said to be complete if no imputed variable in the observation contains soft missing (.). Observations that are not complete are said to be incomplete.

completed-data analysis. The analysis or estimation performed on the made-to-be completed (imputed) data. This term does not refer to analysis or estimation performed on the subset of complete observations.

complete DF, complete degrees of freedom. The degrees of freedom that would have been used for inference if the data were complete.

DA, data augmentation. An MCMC method used for the imputation of missing data.

EM, expectation-maximization algorithm. In the context of MI, an iterative procedure for obtaining maximum likelihood or posterior-mode estimates in the presence of missing data.

flong data. See *style*.

flongsep data. See *style*.

FMI, fraction missing information. The ratio of information lost due to the missing data to the total information that would be present if there were no missing data.

An *equal FMI test* is a test under the assumption that FMIs are equal across parameters.

An *unrestricted FMI test* is a test without the equal FMI assumption.

hard missing and soft missing. A hard missing value is a value of .a, .b, ..., .z in $m=0$ in an imputed variable. Hard missing values are not replaced in $m>0$.

A soft missing value is a value of . in $m=0$ in an imputed variable. If an imputed variable contains soft missing, then that value is eligible to be imputed, and perhaps is imputed, in $m>0$.

Although you can use the terms hard missing and soft missing for passive, regular, and unregistered variables, it has no special significance in terms of how the missing values are treated.

imputed, passive, and regular variables. An imputed variable is a variable that has missing values and for which you have or will have imputations.

A passive variable is a variable that is a function of imputed variables or of other passive variables. A passive variable will have missing values in $m=0$ and varying values for observations in $m>0$.

A regular variable is a variable that is neither imputed nor passive and that has the same values, whether missing or not, in all m.

Imputed, passive, and regular variables can be registered using the `mi register` command; see [MI] **mi set**. You are required to register imputed variables, and we recommend that you register passive variables. Regular variables can also be registered. See *registered and unregistered variables*.

incomplete observations. See *complete observations*.

ineligible missing value. An ineligible missing value is a missing value in a to-be-imputed variable that is due to inability to calculate a result rather than an underlying value being unobserved. For instance, assume that variable `income` had some missing values and so you wish to impute it. Because `income` is skewed, you decide to impute the log of income, and you begin by typing

```
. generate lnincome = log(income)
```

If `income` contained any zero values, the corresponding missing values in `lnincome` would be ineligible missing values. To ensure that values are subsequently imputed correctly, it is of vital importance that any ineligible missing values be recorded as hard missing. You would do that by typing

```
. replace lnincome = .a if lnincome==. & income!=.
```

As an aside, if after imputing `lnincome` using `mi impute` (see [MI] **mi impute**), you wanted to fill in `income`, `income` surprisingly would be a passive variable because `lnincome` is the imputed variable and `income` would be derived from it. You would type

```
. mi register passive income
. mi passive: replace income = cond(lnincome==.a, 0, exp(lnincome))
```

listwise deletion, casewise deletion. Omitting from analysis observations containing missing values.

M, m. M is the number of imputations. m refers to a particular imputation, $m = 1, 2, \ldots, M$. In mi, $m = 0$ is used to refer to the original data, the data containing the missing values. Thus mi data in effect contain $M + 1$ datasets, corresponding to $m = 0, m = 1, \ldots,$ and $m = M$.

MCMC, Markov chain Monte Carlo. A class of methods for simulating random draws from otherwise intractable multivariate distributions. The Markov chain has the desired distribution as its equilibrium distribution.

mi data. Any data that have been `mi set` (see [MI] **mi set**), whether directly by `mi set` or indirectly by `mi import` (see [MI] **mi import**). The mi data might have no imputations (have $M = 0$) and no imputed variables, at least yet, or they might have $M > 0$ and no imputed variables, or vice versa. An mi dataset might have $M > 0$ and imputed variables, but the missing values have not yet been replaced with imputed values. Or mi data might have $M > 0$ and imputed variables and the missing values of the imputed variables filled in with imputed values.

mlong data. See *style*.

monotone-missing pattern, monotone missingness. A special pattern of missing values in which if the variables are ordered from least to most missing, then all observations of a variable contain missing in the observations in which the prior variable contains missing.

original data. Original data are the data as originally collected, with missing values in place. In mi data, the original data are stored in $m = 0$. The original data can be extracted from mi data by using `mi extract`; see [MI] **mi extract**.

passive variable. See *imputed, passive, and regular variables*.

registered and unregistered variables. Variables in `mi` data can be registered as imputed, passive, or regular by using the `mi register` command; see [MI] **mi set**.

You are required to register imputed variables.

You should register passive variables; if your data are style wide, you are required to register them. The `mi passive` command (see [MI] **mi passive**) makes creating passive variables easy, and it automatically registers them for you.

Whether you register regular variables is up to you. Registering them is safer in all styles except wide, where it does not matter. By definition, regular variables should be the same across m. In the long styles, you can unintentionally create variables that vary. If the variable is registered, `mi` will detect and fix your mistakes.

Super-varying variables, which rarely occur and can be stored only in flong and flongsep data, should never be registered.

The registration status of variables is listed by the `mi describe` command; see [MI] **mi describe**.

regular variable. See *imputed, passive, and regular variables*.

relative efficiency. Ratio of variance of a parameter given estimation with finite M to the variance if M were infinite.

RVI, relative variance increase. The increase in variance of a parameter estimate due to nonresponse.

style. Style refers to the format in which the `mi` data are stored. There are four styles: flongsep, flong, mlong, and wide. You can ignore styles, except for making an original selection, because all `mi` commands work regardless of style. You will be able to work more efficiently, however, if you understand the details of the style you are using; see [MI] **styles**. Some tasks are easier in one style than another. You can switch between styles by using the `mi convert` command; see [MI] **mi convert**.

The flongsep style is best avoided unless your data are too big to fit into one of the other styles. In flongsep style, a separate `.dta` set is created for $m = 0$, for $m = 1$, ..., and for $m = M$. Flongsep is best avoided because `mi` commands work more slowly with it.

In all the other styles, the $M + 1$ datasets are stored in one `.dta` file. The other styles are both more convenient and more efficient.

The most easily described of these `.dta` styles is flong; however, flong is also best avoided because the more efficient mlong style is every bit as convenient as flong, and mlong is memorywise more efficient. In flong, each observation in the original data is repeated M times in the `.dta` dataset, once for $m = 1$, again for $m = 2$, and so on. Variable `_mi_m` records m and takes on values 0, 1, 2, ..., M. Within each value of m, variable `_mi_id` takes on values 1, 2, ..., N, and thus connects imputed with original observations.

The mlong style is recommended. It is efficient and easy to use. Mlong is much like flong except that complete observations are not repeated.

Equally recommended is the wide style. In wide, each imputed and passive variable has an additional M variables associated with it, one for the variable's value in $m = 1$, another for its value in $m = 2$, and so on. If an imputed or passive variable is named *vn*, then the values of *vn* in $m = 1$ are stored in variable `_1_vn`; the values for $m = 2$, in `_2_vn`; and so on.

What makes mlong and wide so convenient? In mlong, there is a one-to-one correspondence of your idea of a variable and Stata's idea of a variable—variable *vn* refers to *vn* for all values of m. In wide, there is a one-to-one correspondence of your idea of an observation and Stata's idea—physical observation 5 is observation 5 in all datasets.

Choose the style that matches the problem at hand. If you want to create new variables or modify existing ones, choose mlong. If you want to drop observations or create new ones, choose wide. You can switch styles with the `mi convert` command; see [MI] **mi convert**.

For instance, if you want to create new variable ageXexp equal to age*exp, and if your data are mlong, you can just type `generate ageXexp = age*exp`, and that will work even if age and exp are imputed, passive, or a mix. Theoretically, the right way to do that is to type `mi passive: generate agexExp = age*exp`, but concerning variables, if your data are mlong, you can work the usual Stata way.

If you want to drop observation 20 or drop if sex==2, if your data are wide, you can just type `drop in 20` or `drop if sex==2`. Here the "right" way to do the problem is to type the `drop` command and then remember to type `mi update` so that `mi` can perform whatever machinations are required to carry out the change throughout $m > 0$, but in the wide form, there are no machinations required.

super-varying variables. See *varying and super-varying variables*.

unregistered variables. See *registered and unregistered variables*.

varying and super-varying variables. A variable is said to be varying if its values in the incomplete observations differ across m. Imputed and passive variables are varying. Regular variables are nonvarying. Unregistered variables can be either.

Imputed variables are supposed to vary because their incomplete values are filled in with different imputed values, although an imputed variable can be temporarily nonvarying if you have not imputed its values yet. Similarly, passive variables should vary because they are or will be filled in based on values of varying imputed variables.

A variable is said to be super varying if its values in the *complete* observations differ across m. The existence of super-varying variables is usually an indication of error. It makes no sense for a variable to have different values in, say, $m = 0$ and $m = 2$ in the complete observations—in observations that contain no missing values. That is, it makes no sense unless the values of the variable is a function of the values of other variables across multiple observations. If variable sumx is the sum of x across observations, and if x is imputed, then sumx will differ across m in all observations after the first observation in which x is imputed.

The `mi varying` command will identify varying and super-varying variables, as well as nonvarying imputed and passive variables. [MI] **mi varying** explains how to fix problems when they are due to error.

Some problems that theoretically could arise cannot arise because `mi` will not let them. For instance, an imputed variable could be super varying and that would obviously be a serious error. Or a regular variable could be varying and that, too, would be a serious error. When you register a variable, `mi` fixes any such problems and, from that point on, watches for problems and fixes them as they arise. Use `mi register` to register variables; see [MI] **mi set**. You can perform the checks and fixes at any time by running `mi update`; see [MI] **mi update**. Among other things, `mi update` replaces values of regular variables in $m > 0$ with their values from $m = 0$; it replaces values of imputed variables in $m > 0$ with their nonmissing values from $m = 0$; and it replaces values of passive variables in incomplete observations of $m > 0$ with their $m = 0$ values. `mi update` follows a hands-off policy with respect to unregistered variables.

If you need super-varying variables, use flong or flongsep style and do not register the variable. You must use one of the flong styles because in the wide and mlong styles, there is simply no place to store super-varying values.

wide data. See *style*.

WLF, worst linear function. A linear combination of all parameters being estimated by an iterative procedure that is thought to converge slowly.

Also see

[MI] **intro** — Introduction to mi

Subject and author index

This is the subject and author index for the *Multiple-Imputation Reference Manual*. Readers interested in topics other than multiple imputation should see the combined subject index (and the combined author index) in the *Quick Reference and Index*. The combined index indexes the *Getting Started* manuals, the *User's Guide*, and all the reference manuals except the *Mata Reference Manual*.

Semicolons set off the most important entries from the rest. Sometimes no entry will be set off with semicolons, meaning that all entries are equally important.

A

add, mi subcommand, [MI] **mi add**
Alfaro, R., [MI] **intro**
Allison, P. D., [MI] **intro substantive**, [MI] **mi impute**
analysis step, [MI] **intro substantive**, [MI] **mi estimate**, *also see* estimation
Anderson, T. W., [MI] **intro substantive**
append, mi subcommand, [MI] **mi append**
appending data, [MI] **mi append**
arbitrary pattern of missing values, [MI] **mi impute mvn**, *also see* pattern of missingness
available-case analysis, [MI] **intro substantive**
average RVI, [MI] **Glossary**, [MI] **mi estimate**

B

Barnard, J., [MI] **intro substantive**, [MI] **mi estimate**, [MI] **mi estimate postestimation**, [MI] **mi estimate using**
Bayesian concepts, [MI] **intro substantive**
between-imputation variability, [MI] **mi estimate**
Bibby, J. M., [MI] **mi impute mvn**
binary variable imputation, *see* imputation, binary
Binder, D. A., [MI] **intro substantive**
Boshuizen, H. C., [MI] **intro substantive**, [MI] **mi impute monotone**

C

cancer data, [MI] **mi estimate**
Carlin, J. B., [MI] **intro**, [MI] **intro substantive**, [MI] **mi estimate**, [MI] **mi impute mvn**, [MI] **mi impute regress**
Carpenter, J. R., [MI] **intro**, [MI] **intro substantive**
casewise deletion, *see* listwise deletion
categorical variable imputation, *see* imputation, categorical
certifying mi data are consistent, [MI] **mi update**
Coffey, C., [MI] **intro substantive**
combination step, [MI] **intro substantive**, [MI] **mi estimate**, [MI] **mi estimate postestimation**, [MI] **mi estimate using**
combining data, [MI] **mi add**, [MI] **mi append**, [MI] **mi merge**
complete degrees of freedom for coefficients, [MI] **Glossary**, [MI] **mi estimate**
complete observations, [MI] **Glossary**
completed-data analysis, [MI] **Glossary**, [MI] **intro substantive**, [MI] **mi estimate**
continuous variable imputation, *see* imputation, continuous
convert, mi subcommand, [MI] **mi convert**
converting between styles, [MI] **mi convert**
copy, mi subcommand, [MI] **mi copy**, [MI] **styles**
custom prediction equations, [MI] **mi impute monotone**

D

DA, *see* data augmentation
data,
    exporting, *see* exporting
    flong, *see* flong
    flongsep, *see* flongsep
    importing, *see* importing
    mlong, *see* mlong
    wide, *see* wide
data augmentation, [MI] **Glossary**, [MI] **mi impute**, [MI] **mi impute mvn**
data management, [MI] **mi add**, [MI] **mi append**, [MI] **mi expand**, [MI] **mi extract**, [MI] **mi merge**, [MI] **mi rename**, [MI] **mi replace0**, [MI] **mi reset**, [MI] **mi reshape**
degrees of freedom, [MI] **mi estimate**
    complete, *see* complete degrees of freedom for coefficients, *also see* estimation
Dempster, A. P., [MI] **intro substantive**, [MI] **mi impute mvn**
describe, mi subcommand, [MI] **mi describe**
describing mi data, [MI] **mi describe**

E

egen command, [MI] **mi passive**, [MI] **mi xeq**
EM, [MI] **Glossary**, [MI] **mi impute mvn**
    parameter trace files, [MI] **mi ptrace**
ensuring mi data are consistent, [MI] **mi update**
equal FMI test, [MI] **Glossary**, [MI] **mi estimate**, [MI] **mi estimate postestimation**
erase, mi subcommand, [MI] **mi erase**, [MI] **styles**
estimate, mi subcommand, [MI] **mi estimate**, [MI] **mi estimate using**
estimation
    allowed estimation commands, [MI] **estimation**
    degrees of freedom for coefficients, [MI] **mi estimate**
    posting VCE, [MI] **mi estimate**
    tests after, [MI] **mi estimate**, [MI] **mi estimate postestimation**
expand, mi subcommand, [MI] **mi expand**
expand for mi data, [MI] **mi expand**
expectation-maximization algorithm, *see* EM
export, mi subcommand, [MI] **mi export**, [MI] **mi export ice**, [MI] **mi export nhanes1**

exporting data, [MI] **mi export**, [MI] **mi export ice**, [MI] **mi export nhanes1**
`extract`, mi subcommand, [MI] **mi extract**, [MI] **mi replace0**
extracting $m=\#$ data from mi data, [MI] **mi extract**, [MI] **mi select**
extracting original data from mi data, [MI] **mi extract**
Ezzati-Rice, T. M., [MI] **intro substantive**

F

flong
    data style, [MI] **Glossary**, [MI] **styles**
    technical description, [MI] **technical**
flongsep
    data style, [MI] **Glossary**, [MI] **mi xeq**, [MI] **styles**
    estimating memory requirements, [MI] **mi convert**
    style, [MI] **mi copy**, [MI] **mi erase**
    technical description, [MI] **technical**
FMI, *see* fraction missing information
fraction missing information, [MI] **Glossary**, [MI] **mi estimate**, [MI] **mi estimate postestimation**
frequentist concepts, [MI] **intro substantive**
`fvset`, mi subcommand, [MI] **mi XXXset**
`fvset` command for mi data, [MI] **mi XXXset**

G

Galati, J. C., [MI] **intro**, [MI] **intro substantive**
Gelman, A., [MI] **intro substantive**, [MI] **mi impute mvn**, [MI] **mi impute regress**
`generate` command, [MI] **mi passive**, [MI] **mi xeq**
Graham, J. W., [MI] **intro substantive**, [MI] **mi impute**
Greenwood, P., [MI] **intro substantive**

H

hard missing value, [MI] **Glossary**
Hartley, H. O., [MI] **intro substantive**, [MI] **mi impute**
heart attack data, [MI] **intro substantive**, [MI] **mi estimate**, [MI] **mi impute logit**, [MI] **mi impute mlogit**, [MI] **mi impute monotone**, [MI] **mi impute mvn**, [MI] **mi impute ologit**, [MI] **mi impute pmm**, [MI] **mi impute regress**
Hocking, R. R., [MI] **intro substantive**
home resale-price data, [MI] **mi estimate**, [MI] **mi estimate postestimation**, [MI] **mi estimate using**
Horton, N. J., [MI] **intro substantive**

I

`ice` command, [MI] **mi export ice**, [MI] **mi import ice**
ignorable missing-data mechanism, [MI] **intro substantive**
`import`, mi subcommand, [MI] **mi import**, [MI] **mi import flong**, [MI] **mi import flongsep**, [MI] **mi import ice**, [MI] **mi import nhanes1**, [MI] **mi import wide**

importing data, [MI] **mi import**, [MI] **mi import flong**, [MI] **mi import flongsep**, [MI] **mi import ice**, [MI] **mi import nhanes1**, [MI] **mi import wide**
imputation,
    binary, [MI] **mi impute logit**
    categorical, [MI] **mi impute mlogit**, [MI] **mi impute ologit**
    continuous, [MI] **mi impute pmm**, [MI] **mi impute regress**
    linear regression, [MI] **mi impute regress**
    logistic regression, [MI] **mi impute logit**
    modeling, [MI] **mi impute**
    monotone, [MI] **mi impute**, [MI] **mi impute monotone**
    multinomial logistic regression, [MI] **mi impute mlogit**
    multiple, [MI] **intro substantive**
    multivariate, [MI] **mi impute monotone**, [MI] **mi impute mvn**
    multivariate monotone, [MI] **mi impute**, [MI] **mi impute logit**, [MI] **mi impute mlogit**, [MI] **mi impute ologit**, [MI] **mi impute pmm**, [MI] **mi impute regress**
    multivariate normal, [MI] **mi impute mvn**
    on subsamples, [MI] **mi impute**
    ordered logistic regression, [MI] **mi impute ologit**
    passive variables, [MI] **mi impute regress**
    predictive mean matching, [MI] **mi impute**, [MI] **mi impute pmm**
    regression, [MI] **mi impute**, [MI] **mi impute regress**
    semiparameteric, [MI] **mi impute pmm**
    step, [MI] **intro substantive**, [MI] **mi estimate**
    transformations, [MI] **mi impute**
    univariate, [MI] **mi impute logit**, [MI] **mi impute mlogit**, [MI] **mi impute ologit**, [MI] **mi impute pmm**, [MI] **mi impute regress**
imputation method, [MI] **mi impute**, [MI] **mi impute logit**, [MI] **mi impute mlogit**, [MI] **mi impute ologit**, [MI] **mi impute pmm**, [MI] **mi impute regress**
    proper, [MI] **intro substantive**, [MI] **mi impute monotone**, [MI] **mi impute mvn**
imputations, recommended number of, [MI] **intro substantive**
`impute`, mi subcommand, [MI] **mi impute**, [MI] **mi impute logit**, [MI] **mi impute mlogit**, [MI] **mi impute monotone**, [MI] **mi impute mvn**, [MI] **mi impute ologit**, [MI] **mi impute pmm**, [MI] **mi impute regress**
imputed variables, *see* variables, imputed
incomplete observations, [MI] **Glossary**
ineligible missing values, [MI] **Glossary**

J

Jeffreys noninformative prior, [MI] **mi impute mvn**
Johnson, W., [MI] **intro substantive**

K

Kent, J. T., [MI] **mi impute mvn**
Kenward, M. G., [MI] **intro substantive**
Khare, M., [MI] **intro substantive**
Kleinman, K. P., [MI] **intro substantive**
Knook, D. L., [MI] **intro substantive**, [MI] **mi impute monotone**

L

Laird, N. M., [MI] **intro substantive**, [MI] **mi impute mvn**
Lepkowski, J. M., [MI] **intro substantive**, [MI] **mi impute**, [MI] **mi impute logit**, [MI] **mi impute mlogit**, [MI] **mi impute monotone**, [MI] **mi impute ologit**
Li, K.-H., [MI] **intro substantive**, [MI] **mi estimate**, [MI] **mi estimate postestimation**, [MI] **mi estimate using**, [MI] **mi impute mvn**
Li, N., [MI] **intro substantive**
linear regression imputation, *see* imputation, regression
linear tests, *see* estimation, tests after
Lipsitz, S. R., [MI] **intro substantive**
listwise deletion, [MI] **Glossary**, [MI] **intro substantive**, [MI] **mi estimate**
Little, R. J. A., [MI] **intro substantive**, [MI] **mi impute mvn**, [MI] **mi impute pmm**
logistic regression imputation, *see* imputation, logistic regression

M

M, [MI] **Glossary**, [MI] **mi impute**
    size recommendations, [MI] **intro substantive**
m, [MI] **Glossary**
MAR, [MI] **intro substantive**
Mardia, K. V., [MI] **mi impute mvn**
Markov chain Monte Carlo, *see* MCMC
MCAR, [MI] **intro substantive**
MCMC, [MI] **Glossary**, [MI] **mi impute**, [MI] **mi impute mvn**
    parameter trace files, [MI] **mi ptrace**
memory requirements, estimating for flongsep, [MI] **mi convert**
Meng, X.-L., [MI] **intro substantive**, [MI] **mi estimate**, [MI] **mi estimate postestimation**, [MI] **mi estimate using**
merge, mi subcommand, [MI] **mi merge**
merging data, [MI] **mi merge**
mi
    add command, [MI] **mi add**
    append command, [MI] **mi append**
    convert command, [MI] **mi convert**
    copy command, [MI] **mi copy**, [MI] **styles**
    describe command, [MI] **mi describe**
    erase command, [MI] **mi erase**, [MI] **styles**
    estimate command, [MI] **mi estimate**, [MI] **mi estimate postestimation**, [MI] **mi estimate using**

mi, *continued*
    estimate postestimation, [MI] **mi estimate postestimation**
    expand command, [MI] **mi expand**
    export command, [MI] **mi export**, [MI] **mi export ice**, [MI] **mi export nhanes1**
    extract command, [MI] **mi extract**, [MI] **mi replace0**
    fvset command, [MI] **mi XXXset**
    import command, [MI] **mi import**, [MI] **mi import flong**, [MI] **mi import flongsep**, [MI] **mi import ice**, [MI] **mi import nhanes1**, [MI] **mi import wide**
    impute command, [MI] **mi impute**, [MI] **mi impute logit**, [MI] **mi impute mlogit**, [MI] **mi impute monotone**, [MI] **mi impute mvn**, [MI] **mi impute ologit**, [MI] **mi impute pmm**, [MI] **mi impute regress**
    merge command, [MI] **mi merge**
    misstable command, [MI] **mi misstable**
    passive command, [MI] **mi passive**
    ptrace command, [MI] **mi ptrace**
    query command, [MI] **mi describe**
    register command, [MI] **mi set**
    rename command, [MI] **mi rename**
    replace0 command, [MI] **mi replace0**
    reset command, [MI] **mi reset**
    reshape command, [MI] **mi reshape**
    select command, [MI] **mi select**
    set command, [MI] **mi set**
    st command, [MI] **mi XXXset**
    stjoin command, [MI] **mi stsplit**
    streset command, [MI] **mi XXXset**
    stset command, [MI] **mi XXXset**
    stsplit command, [MI] **mi stsplit**
    svyset command, [MI] **mi XXXset**
    test command, [MI] **mi estimate postestimation**
    testtransform command, [MI] **mi estimate postestimation**
    tsset command, [MI] **mi XXXset**
    unregister command, [MI] **mi set**
    unset command, [MI] **mi set**
    update command, [MI] **mi update**, [MI] **noupdate option**
    varying command, [MI] **mi varying**
    xeq command, [MI] **mi xeq**
    xtset command, [MI] **mi XXXset**
mi command, [MI] **intro**, [MI] **styles**, [MI] **workflow**
mi data, [MI] **Glossary**
missing at random, *see* MAR
missing completely at random, *see* MCAR
missing data, [MI] **intro substantive**
    arbitrary pattern, [MI] **Glossary**, [MI] **intro substantive**, [MI] **mi impute mvn**
    monotone pattern, [MI] **Glossary**, [MI] **intro substantive**, [MI] **mi impute monotone**, [MI] **mi impute mvn**
missing not at random, *see* MNAR

missing values,
    hard and soft, [MI] **Glossary**
    ineligible, [MI] **Glossary**
    pattern of, [MI] **mi misstable**
missingness, pattern, *see* pattern of missingness
misstable, mi subcommand, [MI] **mi misstable**
misstable for mi data, [MI] **mi misstable**
mlong
    data style, [MI] **Glossary**, [MI] **styles**
    technical description, [MI] **technical**
MNAR, [MI] **intro substantive**
monotone imputation, *see* imputation, monotone
monotonicity, *see* pattern of missingness
multinomial logistic regression, imputation, *see* imputation, multinomial logistic regression
multiple imputation, [MI] **intro**, [MI] **intro substantive**, [MI] **styles**, [MI] **workflow**
    analysis step, [MI] **intro substantive**, [MI] **mi estimate**, [MI] **mi estimate postestimation**, [MI] **mi estimate using**
    estimation, [MI] **estimation**
    imputation step, [MI] **intro substantive**, [MI] **mi impute**
    inference, [MI] **intro substantive**
    pooling step, [MI] **intro substantive**, [MI] **mi estimate**, [MI] **mi estimate using**
    theory, [MI] **intro substantive**
multivariate imputation, *see* imputation, multivariate
multivariate logistic variable imputation, *see* imputation, multivariate logistic
multivariate normal imputation, *see* imputation, multivariate normal
multivariate regression imputation, *see* imputation, multivariate
MVN imputation, *see* imputation, multivariate normal

N

nearest neighbors, [MI] **mi impute pmm**
NHANES data, [MI] **mi export nhanes1**, [MI] **mi import nhanes1**
nonlinear tests, *see* estimation, tests after
noupdate option, [MI] **noupdate option**

O

observations, complete and incomplete, [MI] **Glossary**
Olsen, M. K., [MI] **intro substantive**
ordered logistic regression imputation, *see* imputation, ordered logistic regression

P

parameter trace files, [MI] **mi impute mvn**, [MI] **mi ptrace**
passive, mi subcommand, [MI] **mi passive**
passive variables, *see* variables, passive

pattern of missingness, [MI] **Glossary**, [MI] **intro substantive**, [MI] **mi impute**, [MI] **mi misstable**
PMM imputation, *see* imputation, predictive mean matching
pooling step, [MI] **intro substantive**, [MI] **mi estimate**, [MI] **mi estimate postestimation**, [MI] **mi estimate using**
postestimation, *see* estimation, tests after
predictive mean matching imputation, *see* imputation, predictive mean matching
programmer's commands and utilities, [MI] **mi select**, [MI] **styles**, [MI] **technical**
proper imputation method, [MI] **intro substantive**
ptrace, mi subcommand, [MI] **mi ptrace**
.ptrace files, [MI] **mi impute mvn**, [MI] **mi ptrace**

Q

query, mi subcommand, [MI] **mi describe**

R

Raghunathan, T. E., [MI] **intro substantive**, [MI] **mi estimate**, [MI] **mi estimate postestimation**, [MI] **mi estimate using**, [MI] **mi impute**, [MI] **mi impute logit**, [MI] **mi impute mlogit**, [MI] **mi impute monotone**, [MI] **mi impute ologit**
random-number seed, [MI] **mi impute**
register, mi subcommand, [MI] **mi set**
registered variables, *see* variables, registered
regular variables, *see* variables, regular
Reiter, J. P., [MI] **intro**, [MI] **intro substantive**, [MI] **mi estimate**, [MI] **mi estimate postestimation**, [MI] **mi estimate using**
relative efficiency, [MI] **Glossary**, [MI] **mi estimate**
relative variance increase, [MI] **Glossary**, [MI] **mi estimate**
rename, mi subcommand, [MI] **mi rename**
rename for mi data, [MI] **mi rename**
renaming variables, [MI] **mi rename**
replace command, [MI] **mi passive**, [MI] **mi xeq**
replace0, mi subcommand, [MI] **mi replace0**
reset, mi subcommand, [MI] **mi reset**
reshape, mi subcommand, [MI] **mi reshape**
reshape for mi data, [MI] **mi reshape**
ridge prior, [MI] **mi impute mvn**
Royston, P., [MI] **intro**, [MI] **intro substantive**, [MI] **mi estimate**, [MI] **mi export**, [MI] **mi export ice**, [MI] **mi import**, [MI] **mi import ice**, [MI] **mi impute**, [MI] **mi impute monotone**
Rubin, D. B., [MI] **intro substantive**, [MI] **mi estimate**, [MI] **mi estimate postestimation**, [MI] **mi estimate using**, [MI] **mi impute**, [MI] **mi impute logit**, [MI] **mi impute monotone**, [MI] **mi impute mvn**, [MI] **mi impute pmm**, [MI] **mi impute regress**

Rubin's combination rules, [MI] **mi estimate**, [MI] **mi estimate using**
RVI, *see* relative variance increase

S

Schafer, J. L., [MI] **intro substantive**, [MI] **mi estimate**, [MI] **mi impute**, [MI] **mi impute monotone**, [MI] **mi impute mvn**
Schenker, N., [MI] **intro substantive**, [MI] **mi impute**, [MI] **mi impute pmm**, [MI] **mi impute regress**
select, mi subcommand, [MI] **mi select**
semiparametric imputation method, *see* imputation, predictive mean matching
sequential imputation, [MI] **mi impute**
set, mi subcommand, [MI] **mi set**
setting M, [MI] **mi add**, [MI] **mi set**
setting mi data, [MI] **mi set**
single-imputation methods, [MI] **intro substantive**
soft missing value, [MI] **Glossary**
Solenberger, P., [MI] **intro substantive**, [MI] **mi impute**, [MI] **mi impute logit**, [MI] **mi impute mlogit**, [MI] **mi impute monotone**, [MI] **mi impute ologit**
st, mi subcommand, [MI] **mi XXXset**
st commands for mi data, [MI] **mi stsplit**, [MI] **mi XXXset**
.ster files, [MI] **mi estimate**, [MI] **mi estimate using**
Stern, H. S., [MI] **intro substantive**, [MI] **mi impute mvn**, [MI] **mi impute regress**
Sterne, J. A. C., [MI] **intro**
stjoin, mi subcommand, [MI] **mi stsplit**
stjoin for mi data, [MI] **mi stsplit**
streset, mi subcommand, [MI] **mi XXXset**
streset command for mi data, [MI] **mi XXXset**
stset, mi subcommand, [MI] **mi XXXset**
stset command for mi data, [MI] **mi XXXset**
stsplit, mi subcommand, [MI] **mi stsplit**
stsplit for mi data, [MI] **mi stsplit**
style, [MI] **Glossary**, [MI] **mi convert**, [MI] **styles**
style,
    flong, *see* flong
    flongsep, *see* flongsep
    mlong, *see* mlong
    wide, *see* wide
Sun, W., [MI] **intro substantive**
super-varying variables, [MI] **Glossary**, [MI] **mi varying**
survey data, [MI] **intro substantive**
svyset, mi subcommand, [MI] **mi XXXset**
svyset command for mi data, [MI] **mi XXXset**
switching styles, [MI] **mi convert**

T

Tanner, M. A., [MI] **intro substantive**, [MI] **mi impute mvn**

Taylor, J. M. G., [MI] **intro substantive**, [MI] **mi impute**, [MI] **mi impute pmm**, [MI] **mi impute regress**
tests after estimation, *see* estimation, tests after
transformed coefficients, [MI] **mi estimate**, [MI] **mi estimate postestimation**, [MI] **mi estimate using**
tsset, mi subcommand, [MI] **mi XXXset**
tsset command for mi data, [MI] **mi XXXset**

U

uniform prior, [MI] **mi impute mvn**
univariate imputation, *see* imputation, univariate
unregister, mi subcommand, [MI] **mi set**
unregistered variables, *see* variables, unregistered
unrestricted FMI test, [MI] **Glossary**, [MI] **mi estimate**, [MI] **mi estimate postestimation**
update, mi subcommand, [MI] **mi update**, [MI] **noupdate option**
utility routines, [MI] **technical**

V

van Buuren, S., [MI] **intro substantive**, [MI] **mi impute**, [MI] **mi impute logit**, [MI] **mi impute mlogit**, [MI] **mi impute monotone**, [MI] **mi impute ologit**
Van Hoewyk, J., [MI] **intro substantive**, [MI] **mi impute**, [MI] **mi impute logit**, [MI] **mi impute mlogit**, [MI] **mi impute monotone**, [MI] **mi impute ologit**
variables,
    imputed, [MI] **Glossary**, [MI] **mi rename**, [MI] **mi reset**, [MI] **mi set**
    passive, [MI] **Glossary**, [MI] **mi passive**, [MI] **mi rename**, [MI] **mi reset**, [MI] **mi set**, [MI] **mi xeq**
    registered, [MI] **Glossary**, [MI] **mi rename**, [MI] **mi set**
    regular, [MI] **Glossary**, [MI] **mi rename**, [MI] **mi set**
    unregistered, [MI] **Glossary**, [MI] **mi rename**, [MI] **mi set**
    varying and super varying, [MI] **Glossary**, [MI] **mi set**, [MI] **mi varying**
varying, mi subcommand, [MI] **mi varying**
varying
    estimation sample, [MI] **mi estimate**
    variables, *see* variables, varying
verifying mi data are consistent, [MI] **mi update**

W

White, I. R., [MI] **intro**, [MI] **intro substantive**, [MI] **mi estimate**
wide
    data style, [MI] **Glossary**, [MI] **styles**
    technical description, [MI] **technical**
within-imputation variability, [MI] **mi estimate**

WLF, *see* worst-linear function
Wong, W. H., [MI] **intro substantive**, [MI] **mi impute mvn**
workflow, [MI] **workflow**
worst linear function, [MI] **Glossary**, [MI] **mi impute mvn**

X

xeq, mi subcommand, [MI] **mi xeq**
xtset, mi subcommand, [MI] **mi XXXset**
xtset command for mi data, [MI] **mi XXXset**
xxxset, programming, [MI] **technical**